Kit Houses by Mail

Kit Houses by Mail

Brad McDole and Chris Jerome

A Stonesong Press Book

published by

Hearst Books, New York

Designed by Jacqueline Schuman

Cover design by Colette Gaiter

Revised edition prepared by Linnea Leedham Ochs and Peter J. Ochs II

Library of Congress Cataloging in Publication Data

McDole, Brad.
    Kit houses by mail.

    ''A Stonesong Press book.''
    Includes index.
    1. Prefabricated houses. 2. Architecture,
Domestic—Designs and plans. 3. House construction.
I. Jerome, Chris. II. Title.
TH4819.P7M32 1982 691'.97'029473 82-2887
ISBN 0-87851-153-9   AACR2
ISBN 0-87851-152-0 (pbk.)

10 9 8 7 6 5 4 3 2 1

To the Mr. Blandings in every one of us

I prefer houses to the open air. In a house we all feel of the proper proportions. Everything is subordinated to us, fashioned for our own use and pleasure.

— Oscar Wilde

The house shows the owner.

— George Herbert

Contents

The dream can start at any time and for a variety of reasons. You have tired of apartment life, and desire the spaciousness, and the sense of *owning,* that you can get only with a house. You feel the stirring of the craftsman in you, and you wish to make something—something important—that is all your own. You are seeking an alternative, a place in the mountains or on a lake or by the seashore, and you can see every inch of it, every wall, board, and shingle of it, in your mind's eye.

And no matter why you've had the dream, you've decided to make it come true.

The alternatives to pursue are many. You can begin to make the rounds of real estate agents, searching with high hopes for an old house that just happens to have been built for someone whose needs and wishes matched your own, or a new house put up by a builder or developer who just happened to have you in mind. You can contract with an architect and a builder to fulfill your dream, and keep your fingers crossed that the estimate will not be exceeded, and that what appeared on paper will magically reappear on your plot of land. You can try it yourself, arrange for a foundation to be dug, schedule the electricians and the plumbers, pick up hammer and saw—and hope that the joists will hold up the floor, the lumberman will provide the right materials, and your marriage will stay intact.

Or—and this is the reason for this book—you can make arrangements with any of the large number of American house kit manufacturers and enjoy all the advantages they can provide. You will know exactly what the house will look like, before construction begins; you can adjust costs—accurately—to fit your budget; you can do as little, or as much, work as you are willing to do or want to do; and best of all, you can pick a house that fits your land, your use, your family size, and your design ideas almost exactly.

The house kit business in America is a big business, and it is a growing one. Manufacturers range from retailing giants like The Wickes Corporation (through its Wickes Components division) to one-person operations. The range of sizes, styles, prices, financing plans, materials, and options available is equally large.

Kit Houses by Mail is designed to explain what is available, what to look for, what to look out for; to tell you how to determine your own role in the construction of a house from a kit; to advise how (and whether) to bring in a contractor; and to suggest the best, and least expensive, ways to get the work done. It will help you to choose and evaluate a house kit, to plan construction, to be aware of your various rights and how to assert them, and to protect yourself financially without overextending yourself physically. And, in the heart of the book, more than 50 different houses available in kit form—offered by nearly as many different manufacturers—are pictured, described, and priced.

We believe that with this book in hand, you can make intelligent, informed decisions about houses that come in kit form. We hope that if you *do* decide to build your dream house this way, every step along the way will be as pleasurable, simple, and rewarding as possible—right up to the day when you find yourself sitting on the deck of your own specially tailored house, a dream come true.

1
THE KIT ALTERNATIVE

Frank Watson watched his house arrive on the back of two flatbed trucks. If you drive on interstate highways, you already have an image of what he bought. Modular housing. Two of those half-shells that ride assembled down the road, looking as if they had been split by a chain saw and could be reunited with a handful of bolts. But Watson didn't buy modular housing, so his project was slightly less finished when it got packed up for delivery. In fact, it looked like a pile of lumber.

The truck drivers and a few friends helped Frank stack his house on the ground, and he was left with walls, windows, doors, beams, and roof decking; a set of plans; and the prospect of putting together the 40-ton equivalent of a Christmas toy by following the easy-to-read instructions. Watson had bought the ultimate in hobby kits that can be sent to a home—a kit that turns out to *be* the home.

It may seem like an unlikely way to put a roof over one's head, but Watson had surveyed the other options. Recent cost increases on building materials had made stick building, as regular frame construction is called, about as affordable as a hospital stay and as hard to budget as a defense contract. Watson wanted to bypass the expensive architects, the contractor markup, the delays, the lumber shortages, and the erroneous cost estimates that can put a sag in the dream house.

Price was not the only concern. If it had been, Watson might have found a home from among the large variety of spartan prebuilts or manufactured modules coming out of factories all over the country. But Watson also wanted quality, the kind of craftsmanship often left behind as builders standardize techniques and eliminate flourishes that create carpenter overtime. Watson figured that the best way to ensure quality was to do some of his own work. He was not a professional carpenter, but he knew how to drive a nail and chisel a mortise. He thought that a careful amateur builder with a personal stake in the result might do better than a distracted expert with one eye on the circular saw and the other on the pay voucher.

One solution would have been to build an entire house from scratch. Watson did not have enough time for that. A house kit seemed like a sensible alternative. The kit would include precut timbers and preassembled walls, a head start from the factory that would lessen the time it took to put up the shell of the house. Once the house was "dried in" — protected from the weather — Watson could proceed at a more leisurely pace. He could give over his weekends to interior walls, floors, and cabinets — the finish work that shows. He could hire his own people to do the plumbing and electrical work, thus cutting out the contractor and the inevitable markup. He could save some money, learn some carpentry, and take a real part in the creation of his home. He would never suffer from the helpless feeling of the owner-to-be who is shown through his nest-in-progress from time to time but lacks both control over and understanding of the construction.

The concept of producing parts of a house in a factory and then assembling the parts at a building site is growing in appeal. The National Association of Home Manufacturers estimates that more than 200,000 panelized and precut houses were sold in 1981, and this figure does not include modular houses. Of the various types of manufactured housing firms, only a small number sell house kits directly to consumers, and these are the companies we have tried to identify in this book. We contacted all the companies we knew about, and the most interesting, most proficient, and most economical of those that responded are included.

Since Watson was going to build in a rural area in Massachusetts, he chose to deal with a Massachusetts kit company called Habitat, an outfit that specializes in modern glassy designs with heavy beam construction and exposed cathedral ceilings. A Habitat house is a far cry from a plywood box—which makes a point about the diversity of companies that sell house kits. All the basic varieties of stick-built houses can be found in kit form. At one end of the spectrum are simple vacation homes that can be carried into the building site and erected in one or two days. Then there are the low-cost, unadorned houses sold by national companies like Miles or Wickes. Their main virtue is economy, and a few of the kit companies even provide financing plans that require no down payment. A family can contribute its own work, or "sweat equity," to reduce the size of the mortgage. Farther up the price scale, there are split-level kits, colonial kits, ranch-house kits, barn kits, and elaborate post-and-beam structures that feature expensive wood and flexible floor plans and that can cost over $100,000 when completed.

Beyond the traditional dwellings, there are kit houses that you can't get from the run-of-the-mill stick builder. Geodesic-dome kits are relatively easy to put together and offer a lot of square footage at a low cost. (Sometimes a sympathetic builder who scratches his head in bewilderment at a *plan* for a dome will agree to build from a manufactured kit, for which instructions are more precise and some materials are prepared in advance.) Some manufacturers make use of solar energy. Others have special features that would be hard to duplicate at any price. Yankee Barn, for instance, uses old barn timbers in its exposed-ceiling design. Bow House produces curved rafters that make the ceiling of the house look like the inside of an overturned galleon. The curved roof is an old New England idea, developed to shed snow. It was relegated to the archives by modern builders, who found it impractical to bend large pieces of wood, but Bow House can make the rafters in a factory at reasonable cost.

Another old design revived by the kit companies is the log cabin. Making a log cabin from raw timbers can be a tendon-popping experience with appeal only to people who like sweat and chain-saw oil. Making a log cabin from a kit of precut, prepeeled, and prenumbered logs is still no vacation, but the prospect has attracted hundreds of do-it-yourself builders. According to a November 1980 *Wall Street Journal* report, the log-cabin kit business is booming. Since the logs themselves replace four elements in the conventional house—the outer walls, the studs, the insulation, and the inner walls—log construction can cut conventional building costs by as much as half.

Whatever form they take, house kits involve a partnership between self-reliance and the factory assembly line. It is a combination of opposites. Factories thrive on the economies of monotony, while the do-it-yourself builder usually favors beauty, spontaneity, and the pleasures of handwork. By connecting Mr. Blandings to the assembly line, house kit companies use the advantages of both approaches. Good kit manufacturers employ labor-saving techniques and machine accuracy to create quality at the cutting end, and they encourage customer self-help to create quality and economy at the putting-together end. The combination has appeal for several categories of home buyers.

Some people want to build a home or vacation cabin from start to finish. They choose kits because the kits simplify construction and save time during the early phases of the building process.

Others have limited carpentry experience, but still want to build part of the house. Many kit buyers fall into this category. Some of them choose simple kits, like Shelter-Kit, that can be put together by a novice. Others buy more elaborate kits, hire carpenters or a contractor to build the shell of the house from the kit, and then do the remaining work themselves.

Still others don't want to pick up a hammer, but do want control over how the house is built. A kit allows you to be your own contractor. By supervising the construction and hiring the carpenters, plumbers, electricians, and roofers, you can demand quality because you are paying the bills.

Finally, there are those who want a regular contractor to produce the house from beginning to end, but still want the contractor to build from a kit. Although they are not likely to save any money this way, there are other reasons for such a decision. Kit houses are designed by architects, and many people cannot afford the services of an architect. A kit gives them the option of a personalized design without paying a lot of extra money for it. People can visit a kit house, and decide whether or not they like it, before they put up any money. The ease of construction may make it possible to move in a few weeks earlier with a kit house than with a similar-sized stick-built house. Some people buy kits because they don't want to pay extra rent someplace while work on their new house drags on. Others are attracted by some special feature in a kit house, like the genuine old barn timbers used by Yankee Barn.

Frank Watson fit several of these categories at once. He liked the Habitat design, the people at the company, and the assurance that they would give technical advice during various phases of construction. "It was a fairly easy house to put together," Watson says. "We started the foundation on Mother's Day weekend. Since that wasn't part of the kit, we bought all those materials from the outside. By the first of June, we had gotten to the kit, and the house was closed in by the end of June."

Watson was still working on finishing touches as of that September. He figures that he saved maybe three weeks of labor by building from the kit, and he wiped 20% off the cost of contractor construction by doing his own work and by subcontracting—"subbing out"—various technical jobs. He had expected to save more than that, but the price of materials *not* in the kit had escalated since his original calculations were made. Kit prices are usually not subject to such escalations. In most cases, they are guaranteed from the signing of the contract up until the time the house is delivered. How much money you save with a kit usually depends on how skilled you are at buying the extra materials, and how intelligently you subcontract the other jobs.

A house kit does not eliminate all the little pitfalls and daily dilemmas that accompany any building project. Watson had his share of those. But he is still pleased with his choice. "If I had it to do over again," he says, "I think I'd do it the same way. Maybe with more time and with more help I might do it with sticks, but in the situation I was in, I made the right decision."

Judging by the growth of kit companies, a lot of other people have reached the same conclusion as Frank Watson.

2
WHAT'S IN A KIT?

When you buy a kit package, you are really buying two shortcuts in the building process. The first is that the heavy stuff, such as rafters, posts, and beams, is precut and prenotched. You still have to wrestle the timbers into place. But you don't have to worry about making precision cuts and then raising and lowering 400-pound beams when those cuts turn out wrong. The factory turns out a snug fit, so you can avoid a lot of extra carpentry in the air.

The second is that the wall panels are preassembled. How complete they are depends on the kit. But in most cases, the studs are already cut and nailed together, the sheathing is applied, and doors and windows have been prehung and possibly preinstalled. Fitting doors and windows is usually a job that requires patience and precision. The kit company takes the trouble out of that. Nailing up a stud wall is not particularly difficult, but if you are an amateur it can take up a great deal of time. The wall panels may arrive with insulation and exterior siding in place, which saves even more on-the-scene effort.

Most house kits *don't* include the foundation, subfloor, interior drywall, plumbing, wiring, and the finish floor. The kit company will probably send along the roof decking, which is a help, but somebody still has to install it, apply the insulation, and then put on the shingles. Unless you are buying one of the very simple kits, putting up a precut house still requires quite a bit of carpentry around the edges.

That's why it is important to pick a kit that suits your own level of ability. It is unlikely that a person whose last carpentry effort was a wooden ashtray cut on a lathe in high school shop class will be able to handle most of these kits alone. A few kits do give the amateur a chance to start from ignorance and learn as he or she builds. We will try to take note of these in the "Special Features" section of each home we describe. But most of them demand experience. You can't hang out over the rafters and spike together tongue-and-groove roof decking, or cross-brace the joists, or install the shingles, without some working knowledge or at least some expert advisers. Of course, you may not want to hog the labor anyway. Kit owners have been happy to serve as full-time carpenters, carpenter helpers, contractors, interior finishers, sanders and painters, or just interested bystanders at the construction site. As long as you approach both the kit and your own abilities realistically, you will not be disappointed.

The nice thing is that the kit people will not want to mislead you on this point. Salesmen and dealers will be frank about whether you have the skills, and whether the company is equipped to take on the burden of helping do-it-yourselfers with their problems and answering their questions. If a company is not accustomed to working with amateurs, then the percentage of buyers who build the kits themselves will be low. Ed Sink, of New England Log Homes, reports that 37% of his firm's buyers do the whole thing themselves, 60% hire professionals to erect the shell, and the remainder "turnkey" the job, meaning that they get a contractor to do everything.

Another thing a kit company gives you is a step-by-step instruction book, which should be much closer to reality than a blueprint. If the company has dealt intelligently with do-it-yourselfers, it will produce an easy-to-understand book, and it will have worked the bugs out of the house plans. You can ask to see one of the instruction books. A few companies even send tools along in the kit package, a definite sign that amateur labor is encouraged.

In our talks with kit companies and kit owners, we have discovered three laws of kit building that seem to apply to all kit houses. The first is that the bigger the house, the harder the job. It is exponentially easier to handle the carpentry in a 50-square-foot module than it is to deal with long rafters and huge roof spans. The second law is that the more flexible the floor plan, the more complicated the construction. The rank amateur may be advised to avoid post-and-beam houses or barn kits that advertise a different design for each buyer, unless he or she hires a professional to put up the shell.

The third is that the kit isn't everything. The work done in the factory is not necessarily the most demanding work, from a technical standpoint. If you want to build the whole thing yourself, you need to consider the complexities that both precede and follow the nailing of those prepared stacks of lumber.

As you browse through the book, you will notice that the house kits fall into several specific categories. Kit companies usually produce a variety of designs, some simple and some complex. In our effort to emphasize do-it-yourself possibilities, we have tried to pick the simpler designs offered by each company. Although there are differences between one manufacturer and another, here is a general rundown on the types of houses represented, and the building skills required by each.

Modular, Shedlike Structures

Shelter-Kit and Cluster Shed fit this category. These are the easiest projects for the novice. The components are light, the designs are simple, and the instructions are understandable. The basic units are small, but you can tack a bunch of them together to create a larger building. These kits offer the construction advantages of small spaces plus the possibility of roominess by addition.

Michael Pollet of Washington, D.C., is a typical Shelter Kit owner. He is an architect who wanted to build an addition onto his vacation home in the mountains of Virginia. He knows how to design buildings, but he had never driven his own nails before. Pollet describes the experience:

"We already had a weekend house, but we needed more space. The twelve-by-twelve Shelter-Kit module seemed perfect. We wanted to connect the deck of the house kit with the deck we already had—sort of build a bridge between them.

"They loaded it up at the factory, and I brought it down in a U-Haul. My wife and I were going to put it up ourselves, but my wife's doctor was also interested. He wanted to build some of these kits in Florida, and he helped us with ours. The three of us worked on it. I'm sixty-six, and at the end of the first day I knew I wasn't going to put in any more full days. The three of us did everything but put in the piers for the foundation. It went very smoothly. There were a couple of things they could have made clearer, but they were very minor things. Three weekends pretty well cleared it up."

Domes

Dome kits are among the least difficult to erect. The skeleton of the dome goes up like a huge Tinkertoy set, with metal hubs that join the pieces of wood to form triangular shapes. The Cathedralite company claims that dome shells can be finished in eight hours. One of the few skills required is the ability to tighten bolts.

The hard part about domes is making them leakproof at the seams. There is no protective division between the roof and walls in a dome. In fact, roof and walls are all the same thing. The intersections of all the little triangles must be sealed from the deck to the top of the dome. A few new chemical compounds are being used to seal domes, but the tried-and-true method is asphalt shingles. It takes a certain amount of skill to apply shingles correctly.

The interior work in a dome can also be difficult. You have to cut up a seemingly endless number of little triangles and fit them together over the studs. If you are using drywall, there will be plenty of seams to tape and finish.

Economy Homes

These are the houses with standard truss roofs and simple rectangular shapes sold by companies like Miles and Wickes. These houses are designed for people who might not otherwise be able to afford any home. Miles Homes claims that the company "was literally built on the idea that if you are long on ambition, but short on ready cash, you deserve to be trusted and given the chance to build your own home." Since Miles finances the kits, the company is careful to see that the owners can carry through with the carpentry. They don't want to be stuck with the responsibility of suburban renewal. The company provides plenty of support, easy plans, and explicit building instructions. Miles aims its pitch at people who are "skilled with simple carpenter tools."

Log Cabins

Log homes are popular with do-it-yourself builders, but log homes require muscle. You can't toss around 5-inch-diameter logs, especially if they are 16 feet long. Some builders use cranes to lift the heavy log rafters into place, so it takes some effort to get them up by hand. The logs may be nailed together with very long spikes, which demands considerable heft with a hammer.

The wall logs of a log-cabin kit are prenotched and, frequently, numbered, so they go together like a giant Lincoln Log set. The hard part is getting a tight fit. The logs are grooved, and the grooves must match up perfectly. Since logs form both the inside and outside walls, a good caulking or sealing job is critical.

Some amateur builders have no trouble with the walls, but report complications around the roof and the gable ends of the houses. Log homes are within the technical reach of people who have some strength and some carpentry experience, but not of the complete novice.

Post-and-Beam Houses

These are the barns, deck houses, and expensive designs that feature exposed beams and exposed roof decking. If you have some building experience, as Frank Watson did, you can tackle this kind of house, but if you don't, you are advised to limit your role to that of contractor and inside finisher. Habitat recognized that its houses were not designed for do-it-yourselfers and now offers a simpler package called Timber-Kit. Timber-Kit is far easier than Habitat, but you still have to know some carpentry to nail it up.

3

IS A KIT
RIGHT FOR YOU?

Before you start looking at house kits, there are one or two snags that could keep you from buying any of them. Truck access, for instance. While small kits can be carried to the building site by hand, most kits are delivered to the foundation by one or two large flatbed trucks. If your lot isn't wide enough to accommodate the turning radius of a 40- or 50-foot multiwheeler, you may have to cross house kits off your list. Assembled wall units are too heavy to carry more than a few feet. It's hard enough to lay them on the ground right next to the foundation.

Frank Watson says the arrival of the trucks is an amazing sight. "I had no idea how huge those two trucks were going to be. Just because they can get a cement truck in there, people assume a flatbed can make it. It's not always true. Flatbeds need an awful lot of room to turn around."

The delivery requirement is a good bet for an opening question to a kit company. This is how one company, New England Log Homes, tells its customers about trucks:

> The site must be fairly level and there should be enough space for unloading and sorting logs. The majority of log homes are shipped on two trailers, so there must be plenty of room to maneuver these large vehicles. The roads to the home location must be able to support the weight of 60,000 pounds gross vehicle weight. The site must not be in a muddy, rutted condition. Dunnage [loose boards used to keep materials from making contact with the ground], such as 4×4s, must be supplied by the homeowner for stacking the log packages during unloading . . .

While you are measuring for trucks, a call to the local planning department might also be a good idea. Local builders have to abide by local regulations, so there is built-in conformity. House kits are shipped all over the country, which sometimes creates an acceptance problem. You can check the small details when you actually choose a house kit and a plan, but at this stage, you can at least inquire if the type of home you want is allowed by the local powers. Unusual structures like domes are sometimes rejected out of hand because people don't understand them or don't like the way they look next to more conventional homes.

As long as you have the planners on the line, you might ask if amateur work is allowed. In some urban areas, there are severe constraints placed on people who want to build their own houses. Before you figure out how much money you will save by doing it yourself, find out if they will let you.

Truck turnaround is something you don't usually think about when you are planning to build or buy a house. There are many other considerations that apply only to kit houses, or at least more to kit houses than to stick-built houses. In the following pages, we will focus our attention on those special requirements of kit building, and on the advice of people who have built kits themselves. Our purpose is not to give a general course on how to talk to contractors or how to build a house. It is to help you minimize the difficulties that make the difference between a good kit experience and a bad one.

When you choose a contractor to build your house, you want to know that he does quality work and charges fair prices. A couple of local inquiries can get you that information. But when you choose a house kit, the elements of decision are more complex. Before you put any money down, you will want to know whether the kit is worth the price, whether the instructions and blueprints are easy to understand, whether an amateur can put the pieces together, whether deliveries are made on schedule and materials arrive in good condition, whether the company replaces broken or missing pieces, whether materials are of high quality, whether the house lends itself to reasonable energy costs, and whether the budgets and estimates provided by the company are accurate. These questions cannot be answered without a little detective work, which is part of the challenge and fun of house kits. Contractors deal with such questions all the time, and you become a contractor, in a sense, the minute you start investigating one or more kit companies. The emotional rewards of wielding the hammer and saw will come later, but it's a safe bet that you will save as much or more money by cleverly wielding a more familiar tool—the telephone.

There are three sources of local information about kits and kit companies. If you talk to one or more of them before you make formal contact with a kit company, you will have a long head start in evaluating kits and deciding whether they are worth the money.

Approach an Owner

The best way to answer most questions at once is to contact somebody who has already bought a kit that you like. We can't overemphasize the wisdom of such a contact. Most owners are happy to talk about their debut with the hammer, especially if the house is still standing, and you can learn firsthand what skills are required to tackle various stages of the project. You will pick up a few tips that are left out of instruction books. Several owners, for instance, have told us that unloading the trucks was harder than they thought. After groaning over huge timbers for half a day, they wished they had bought more beer and invited more brawn than the three helpers the kit company recommended.

Owners can also give you critical information on how much money you save by doing how much work. We deal with money in the next section of this book, but there is no better source of real numbers, as opposed to the fantasy numbers that find their way into almost every building budget, than people who have already spent the money.

The impression you get from talking to one or more kit customers may be a better index of satisfaction than the bottom line of the account book. Most kit companies understand this and freely provide lists of people who have bought kits and who are willing to talk about them. Sometimes, owners become part of the dealer network. These people are usually honest about their experience, but it may also be useful to seek out an owner who is not on a list and may have had a less positive reaction to the kit.

Owner contact is important enough that if you are a pioneer, willing to buy a kit from a company expanding into new territory, you are at a significant disadvantage. You may be able to compensate by asking the kit company for a special first-owner discount. You are taking extra risks because you cannot benefit from other people's experience. Some companies will let go of a little profit in return for getting an ally in a new area, especially if you agree to tell other customers about your experience.

Approach a Dealer

The next-best source of information about a house kit is the local franchised dealer, or salesman. A few kit companies do not sell through dealers, but most do. Many companies, in fact, will allow you to buy their kits *only* through a local dealer. These people are usually contractors or builders who work for the kit company part-time. In some cases, dealers only push kits, but an active dealer may play a major advisory role through all phases of the construction. He may survey the original building site to make sure it is suitable for a kit, check the foundation for strength and dimensions, inspect the materials as they are unloaded from the truck, and visit the site periodically to see that the walls are right side up. If you want to be your own contractor, the dealer will probably have the names of local carpenters, plumbers, and electricians that you might hire. If you want an experienced crew to put up the shell of the house, the dealer can probably get you together with one.

The dealer, then, can serve as surrogate contractor. He can make your job a lot less risky. He knows the requirements of the kit, and he knows the local wages, prices, and building regulations. A good dealer can be an invaluable source of support, especially if you have no other expert to consult.

Approach a Contractor

If you plan to do the finish work on the house by yourself, you may want to hire a contractor to build the shell. A contractor may be more expensive than a building crew recommended by the dealer, but it's easy enough to compare prices. A contractor may agree to take the kit to the "dried in" stage, or he may suggest that he could produce a similar shell from sticks for less money. It will be hard to get a definitive estimate at this point, since you can't get detailed floor plans from the kit company until you put some money down. But if you are planning to build one of the standard kit houses, you will be able to find out the basic specifications and dimensions—enough to ask for at least a rough estimate.

You can use a contractor to help decide if the kit is worth the price. If he can build a structure of equal quality from sticks for less money, there may be no point in pursuing the kit. A contractor can also give you some idea of the final price of a completed kit house, as compared to the standard per-square-foot rates in your area.

Economic estimates may not be the whole story. Some contractors will not build shells and abandon them to amateur nail-pounders. Some will. You will have to ask around to find out what is possible in your area.

There are two things to keep in mind when you take your kit project to regular contractors for estimates or advice. The first is to make sure that the contractor estimate involves materials of the same quality as those included in the kit. The kit company might be sending old barn timbers while the contractor is figuring for pine beams, or the kit may use select wood for ceiling decking while the contractor calculates for lower-grade construction wood. There are innumerable items that have cheaper substitutes, which make for faulty cost comparisons. The same advice, as we will see, applies when you evaluate one kit against another.

The second thing to remember is that a contractor might deliberately deflate his price, to lure you into doing business with him. The best remedy is to approach a contractor whom you know to be reliable. Stress that you are seeking an estimate only for comparison purposes. A contractor may charge you a fee for preparing such an estimate, if he is not going to be involved in the job itself.

4
FIGURING THE COST

The appeal of a kit to most buyers is that it might save them money. Just looking at the price lists for the kits can create euphoric expectations. You see an elaborate 2,000-square-foot barn for $20,000, and you figure $10 a square foot, a number that has not been heard around construction sites since the late 1950s. Labor won't add much to that, especially if you bring in the family, friends, and neighbors. The rest of the materials you can get on sale or at the salvage yard. The finished house shouldn't cost more than $30,000, or $35,000 at the most. At today's prices that's a house at a 50% discount.

Such calculations are as common as they are improbable. You can save money with house kits, but it is unlikely that you will save anywhere near 50%. The cost of the kit itself, appealing as it looks alone in a booklet, represents only 25% to 50% of the cost of the finished house. How carefully you can estimate that final cost will determine the extent of your pleasure or your dismay.

All three sources of information described in the preceding chapter will give you some notion of the final cost on the house kit you want. The person who has already built a kit is probably the least self-interested source of cost information. If you can find somebody who has put up the same house *recently,* you can rely on the estimate. We stress "recently" because the prices on building materials have a habit of taking great leaps forward in a very short time. Plywood might cost $12 a sheet this month, and $16 a sheet two months from now. If you need 50 sheets, that difference alone adds up to $200.

There's more to a house budget than many buyers think. Here is a typical cost breakdown for a standard cape-style house produced by Heritage Homes of New England. It was computed in November 1981. Although various kit companies have different ways of figuring such costs, this estimate shows the relative value of the items in the kit and the items not in the kit. The kit itself, including the labor to erect the complete shell, makes up less than 50% of the total budget. In more elaborate houses, the kit represents an even lower percentage of the total.

November 3, 1981

Dear Mr. and Mrs. _____:

Our proposal on the 26 × 38 Cape Cod Model, unfinished down, is as follows:

Heritage material package, sales tax, and complete shell erection	$23,507.00
Our approximate costs on the following subcontract items:	
Complete concrete program, including foundation, hatchway, fireplace, and steps	6,085.00
All electrical work including the service entrance, 60 boxes, $250 fixture allowance, hood fan, bath fan, smoke/fire detectors, dryer and range wiring	1,635.00
Plumbing and oil-hot-water heat, two zoned	4,259.00
Drywall and superinsulation work	2,548.00
Carpeting and inlaid linoleum @ $15 per yard average	1,950.00
Cabinets, vanities, counter tops	1,610.00
Gutter and downspouts	193.00
Combination doors	250.00
	$42,037.00
Value of your inside finish carpentry labor	1,088.00
	$43.125.00
Estimated cost of septic system, water connection, site work	3,000.00
Estimated cost of painting and decorating material	750.00

Your "sweat equity" in painting, labor, seeding lawn,
supervision of subcontractors   2,000.00
                                                                      $48.875.00

Our overhead and profit if we handled the subcontracting
for you   2,917.00
                                                                      $51,792.00

Add your land value   - - - - -

   With these figures, you and your banker can arrive at a proper construction loan figure.

   We will take care of all the necessary plan service, specifications, etc., as soon as we receive your deposit and signature on a purchase agreement.

   Thank you for the opportunity to quote on this project; may we hear from you soon?

Yours very truly,
HERITAGE HOMES OF N.E., INC.

   Heritage provides such detailed cost breakdowns to its customers before they put up money, but Heritage does not like to sell to do-it-yourselfers. There is a bit of a hitch here. If a kit company encourages amateur participation, and sells all over the country, it is less likely to keep up with local labor and materials costs. (An active local dealer can help alleviate this problem.) Most companies that cater to do-it-yourselfers cannot give the kind of firm estimate the Heritage people give. Or even if they can, they will probably not want to do the math until the customer makes a down payment and commits himself to the kit. It is a builder's Catch-22. You can't know the true value of a kit until you know the final price of the completed house. You can't know the final price until you get specific bids from plumbers, electricians, and other subcontractors. You can't get specific bids until you have blueprints from the kit company. You can't get blueprints until you have chosen the kit and paid some money.

Consider the example of Charles Firestone, a tall, bearlike man who makes a living selling computer programs. Firestone decided to put up an elaborate barn kit on some land in Virginia. He wasn't interested in doing much carpentry, but he was excited by the idea of serving as contractor.

Since his barn was a personalized design, the kit company had to draw up special plans. They did not send the final blueprints until 10 days before the kit arrived at the site. By this time, Firestone had already arranged his mortgage with a bank, paid the barn company a hefty deposit, and waved some rough drawings and general specifications in front of a few subcontractors. "The estimates that they laid out were low," Firestone says. "In fact, the estimates were *very* low."

So low, in fact, that Firestone was 100% off on subcontractor budgets. Firestone had thought he could build the kit house for $80,000. At last accounting, he was up to $100,000 and still building. He says the final cost will hover someplace around $110,000.

Even with the overruns—an extreme case, to be sure—Firestone likes his kit house. He attributes part of the cost problems to his insistence on quality. "If I didn't like an inside wall, I made them tear it up and do it again." Part of the problem had to do with the nature of the structure itself, which is all custom-built and "doesn't have a square corner in it." Part had to do with the fact that his local subcontractors had never built this type of home and had underestimated their time factors. "And part of it was my fault. I had never done estimates before."

Firestone knows he could make more careful estimates if he had it to do over again. He talked to only one of each category of worker — one plumber, one electrician, one drywall specialist. He offered all of them cost-plus contracts, which are completely flexible, as opposed to a more confining total-price contract. With cost-plus, the subcontractors did not have any great incentive to produce accurate estimates.

As extreme as it is, Firestone's overrun illustrates the necessity for caution. It is easy to misjudge the final cost on a kit house. Even if the kit company provides early estimates that the cost of their kit represents one-third or one-half the cost of the completed house, such estimates are usually not specific enough to be useful. The kits most susceptible to error are the customized designs that feature flexible floor plans and changeable outside dimensions. If you find a house kit with an appealing design, stick to the basic floor plan as much as possible, especially if you are going to contract the job yourself. You will have a better chance of getting accurate estimates and of keeping costs down.

Contractors and local kit dealers may be helpful in writing up a budget, especially if you go with a basic kit model and already know the dimensions and general specifications of the house. The most useful budget is the one that breaks the job into various categories. There are many ways to do this, but we think the following categories should be considered. The first three steps must be taken before the kit comes into play.

SITE PREPARATION. This includes surveying, grading, septic-tank work, cutting trees, leveling ground, adjusting for drainage, compensating for slope and the presence of rock, etc.

FOUNDATION. Includes digging the hole, concrete pouring, block work, waterproofing, drainage, and backfill.

JOISTS AND SUBFLOOR. Includes attaching a sill to the concrete blocks, nailing together the floor structure, and putting plywood on the joists (the members that support the floor).

SHELL ERECTION. Includes putting up the walls, beams, rafters, and exterior siding, all from the kit.

DRYING IN. Includes adding the shingles, flashing, soffit and fascia (the pieces of siding that cover the joints of roof-and-wall and wall-and-wall), doors, and windows. (In some houses, the windows are inserted before the exterior siding is applied.) All these materials are usually included in a kit.

PLUMBING. Includes roughing in and fixtures.

ELECTRICAL. Includes sizing the service for proper amperage and installing breaker box, outlets and switches, and wiring.

HEATING AND AIR CONDITIONING. Includes ductwork, insulation, and furnace installation.

INTERIOR FINISH. Includes installing finish floor, kitchen cabinets, interior walls, baseboards, and door and window trim, and all interior painting and staining.

EXTERIOR FINISH. Includes painting, caulking, trim work, screens, guttering, landscaping.

Dividing a budget into these categories will enable you to pinpoint areas where you can save money, areas where your own labor can reduce costs. You can add up those estimated savings in an effort to decide if a house kit is worth the money, and the time. The contractor or kit dealer can provide you with one of these itemized budgets, but it might be more worthwhile to seek out specific bids from a subcontractor in each of the separate areas. You can lump all the separate bids together, along with the cost of the kit and the cost of the extra materials you might need, to see if the total amount estimated coincides with the general cost-per-square-foot figures you have been given by the kit company. If all the estimates are reasonably accurate, you have a good idea of the final cost of the kit house. It is also worth estimating the value of your own time, and figuring that in — unless building and contracting are your hobbies.

As a very general rule, kits constructed completely by contractors turn out to be just as expensive as comparable stick houses. When the buyer serves as contractor, he can expect to save between 10% and 15%, if he does the job well. When the buyer serves as worker, he saves the salaries of the people his self-help eliminates. Labor costs usually add up to one-third of the total cost on a finished house.

These figures are so rough as to be almost fanciful, but we provide them as a point of departure before you do your own calculations. When you deal with kit companies, insist on finding accurate ways to estimate the final cost of the kit houses.

5

HOW TO CHOOSE BETWEEN HOUSE KITS

There may be only one house kit company that appeals to you, in which case your choice is reduced to a simple yes or no. But because of the recent growth of the precut-housing business, several kit companies are likely to make the same style of house. There are numerous log-cabin outfits, a few dome companies, a smattering of barns, a couple of vacation modules, and a lot of post-and-beams. How can you decide between them? What are some of the signs of quality? How can you compute the best value for the money?

As you might have guessed, there is no simple formula for all of this. Shopping for best value can, at first, be as mystifying as deciding whether this year's trip to Bimini gave more pleasure per travel dollar than last year's trip to Aspen. Between any two kit packages, the designs will be different, the requirements for construction will be different, the percentage of materials included in the kit will be different. One kit may offer superior wooden windows but rather ordinary plywood exterior siding, while another might offer elegant cedar siding but cheap aluminum windows.

The best we can do is to list the points of comparison so as you read through the descriptions of the kits, you can make your own judgments. Since no single factor will be of equal importance to any two customers, the points below are arranged in random order.

Company Reputation

A lot of people have gone into the kit business, and some of them won't be staying around. Since most companies require a rather hefty percentage of the kit cost as a down payment, it is important to find out if they will remain in business long enough to deliver the product. An established sales record and a few years' experience are good signs of reliability, but with the newer and smaller companies you might want to check credit ratings and bank sources.

Proximity

Working with a nearby kit company has some advantages. When most of a house gets piled onto a truck, there are plenty of chances for parts to be broken or pieces to be left out. House kit companies are usually more than willing to replace forgotten beams or defective windows, but the logistics can get complicated if the company is 1,500 miles away. If the part cannot be replaced locally, your work could be delayed. (Some companies guard against such occurrences by sending extra materials and replacement parts.)

A second benefit of proximity is that it reduces the transportation costs usually paid by the kit buyer. Transportation costs can be significant. New England Log Homes provides its customers with the following estimates of East Coast truck rates:

Mileage	1 Flatbed Truck	1 Boom Truck
1-100	$280	$335
101-200	$335	$415
201-300	$445	$530
301-over	$1.45 per mile	$1.70 per mile

These rates are subject to ICC surcharges and may vary weekly.

Not all house kits need boom trucks, which are only used to unload very heavy materials. But even the flatbeds are expensive enough. When you get beyond a radius of about 300 miles, transportation costs can become a determining factor in whether a house kit is worth the money.

Having a manufacturer close at hand also opens up more opportunities for interchange between the do-it-yourself builder and the in-house experts. When problems arise locally, it is a simple matter for a company to send help. But what happens when a kit owner in Alabama calls the company in Connecticut to complain that the joists don't reach the sill plate?

These potential drawbacks are lessened to some degree by the presence of the local dealer, especially if he becomes an active participant in the kit-building process. The dealer usually understands the kit well enough to correct problems himself, and he may even stockpile some replacement materials, which will solve the delay difficulties. Many companies are also setting up new factories in various parts of the country, so that shipping can be decentralized and transportation costs kept low.

If the house kit that you prefer is not common in your region, a final word of caution is in order. You should make sure that the design of the house fits the climatic requirements in your area. A kit manufactured in Massachusetts will probably offer a roof with little or no overhang, to minimize the snow load and to maximize the amount of sun that reaches the interior of the house. Such a roof structure will not necessarily adapt well to Florida, where large overhangs reduce the need for air conditioning in the warm months. A kit which uses un-treated lumber in the floor-joist system (many kits do not include floor joists) may not be transferrable to an area where preservative-treated lumber is necessary to prevent rot and termite infestation.

Warranties

When you buy a home from a contractor, the work is customarily guaranteed for a period of one year against defects like cracked foundations, leaky roofs, sagging walls, twisted timbers, punctured pipes, or sticky windows. The contractor is responsible for defective materials as well as defective workmanship. Among all the workers, suppliers, and subcontractors, the buck stops with him.

When you buy a kit and let a contractor build the whole thing, you get the same protection. If you play contractor yourself, the question of responsibility gets more confused. It doesn't happen often, but there can be annoying minuets around a liability.

Let's say that a window has been broken during delivery. Who will pay up? The company that made it, the kit company, the trucking company that brought it, or you? A kit company usually guarantees all the lumber and the wall panels that arrive with the kit. But that guarantee does not apply to items which carry their own brand name, like windows or shingles. Those things are protected by the original manufacturer, and that guarantee is passed through the kit company to you. So the kit company may hand the window problem to the window company, which may toss it to the trucking company, which may in turn bump it back to the kit company—a circuit that will take time to complete. Meanwhile, you need a new window.

The ease with which complaints are resolved depends a lot on the attitude of the kit company. It is worth a couple of specific questions to find out how the company views its own responsibility in matters like pieces broken during delivery, or materials damaged because the do-it-yourselfer or the subcontractor did not understand the instructions. Most companies are willing to make good on damaged materials, but it is important to clarify areas in which responsibility is confused.

Limited guarantees are not foolproof, but they are usually all you will get from kit companies not willing to stand completely behind a house built with do-it-yourself labor. A few companies give rather extraordinary guarantees on the entire house, for periods of from one to five years, which means they have great confidence in their kits and in the people who build them. When such unusual guarantees are offered, we will try to mention them in the specifications of each kit described in this book.

Quality and Value

Since house kits are sold across local boundaries, they are usually designed to satisfy the most stringent of the regional building codes. They must be good enough to meet the approval of skeptical local inspectors. They must pass the basic wind, load, stress, and energy tests, at least on paper. Many of them conform to FHA, VA, or Farmer's Home standards so a buyer can get mortgage insurance or financing from those agencies. The lumber that is used in kits is graded—that is, inspected and given a rating. When you buy a kit you are getting a general level of quality that at least equals and may even surpass what you might get from a neighborhood builder.

You don't have to fret about minimal standards. But you still want to be able to judge, between one kit company and another, which offers more quality for less money.

In one sense, high quality is a very subjective matter. Some kit companies spend a lot of money to hire good architects who then produce workable and intelligent designs for the kits. When you buy a kit, you are investing in design as well as in material, and designs cannot be compared at the cash register—not, at least, until you sell the house to someone else.

Since design is a subjective question, you will probably resolve it subjectively. The house either grabs you or it doesn't. It either looks good or it doesn't. All the designs are structurally sound, so you don't have to wonder if the house will fall down. The other two important *practical* ramifications of design are ease of construction and energy efficiency. We have already discussed the former, and energy takes up the next section of this book.

A more objective way of evaluating the kits is to compare materials. You can figure out the amount of stuff you are getting from one company, stack it up against the amount you get from another company, and then look at the kit prices for each. Grocery-cart tactics work very well. If you can estimate accurately how much wood is in the kit— how many studs, how much 2×6 decking, etc.—and if you can guess the amount of insulation, fasteners, windows, doors, and incidental items that are included in the kit, you can take the shopping list down to a local lumber store and ask them to give you a price on the whole package. A less-than-honorable way to do this is to pretend you really want to buy the materials from the lumber store, but it is probably better to be forthright. The lumber company might levy a small charge for the estimating service.

The problem with this approach is that the kit company is unlikely to tell you how much lumber goes into the kit. They will tell you what kind of lumber, and what kind of windows, but they may wiggle and waffle on the question of quantities. When you figure out quantities, you have a good head start on guessing the gross profit margin for the kit itself.

If you know somebody who has built a home, or somebody who is a contractor, he can take a walk through a model kit home, or look at the general specifications and floor plans, and probably figure quantities closely enough. For every 4 feet of walls, you have so many studs, so much paneling, so many nails, and so much insulation. For every 100 square feet of roof decking, you have so many board feet of lumber, so many bundles of shingles, etc.

Once you have a list, the only thing to worry about is whether the lumber company is pricing for the same grade of wood and the same quality of windows, insulation, and doors. It is best to specify brand names, because there are enormous differences in prices of items like doors and windows. The kit company will most likely provide you with the brand names of all the materials they use.

Many benefits can come out of this little bit of snooping. For one thing, the grocery-cart method solves an inherent problem with house kits—they don't all contain the same parts. When you buy a kit from one company, you may get the stud walls, roof decking, structural timbers, exterior sheathing, and exterior siding. Another company's package may include only timbers, studs, and roof decking. It is hard to cross over and make comparisons between different packages. But what you can do is compare the cost of each package at the lumber yard with the price of the kit. Allowing for extra services such as an architect's fee and the ease of buying from one source, you can get a fairly good idea of the kit's *value,* as opposed to its price.

When you have priced a kit, assuming that it contains no special materials that are not available in your locality, you will have a sound basis for determining whether the extra advantages of a kit are worth the extra money you pay the kit company instead of the lumber store. (We're assuming that the kit package costs more than the same stuff bought at the lumber store.) If a kit company charges $10,000 for items that can be purchased locally for $5,000, then you have to decide whether the do-it-yourself features and the factory precutting are worth more than $5,000. If one kit company offers $5,000 worth of materials for $8,500 and another sells $4,000 worth of materials for $8,000, then all other things being equal, the first kit is the better buy.

Item-by-Item Evaluation

Maybe you don't want to go to the trouble of trying to figure out how many studs and how much plywood are in a prefabricated kit wall. But you still can't decide between two or three kits of similar design. There is one other comparative method, which isn't nearly as **definitive** as the shopping-cart tactic. But it does take less effort and time. You take the various components that go into each house and evaluate them, one-on-one. It's like figuring out who will win a football game by rating each player against the player at the same position on the opposing team.

No house kit is going to be strongest at all positions. But by going through the basic components, you might get a feel for which kit is best overall—relative to the price. You must be careful with this kind of evaluation, though. A house kit may use cheap aluminum windows and common plywood siding on the outside, and contain beautiful old barn timbers and cedar roof decking inside. No single component can be good enough or bad enough to make or break a kit house. Component comparison is not even remotely an exact science. But as you trace the kit from the ground up, you may get a sense of which company cares the most about quality.

JOISTS. Joists are the pieces of lumber that form the superstructure underneath the floor and distribute the weight of the house onto the sill and the foundation. Many house kits do not include joists at all; you are expected to build your own foundation and subfloor before the kit arrives. If this is the case with your kit, forget about joists at this point.

If a kit does include joists, the thing to note is their width. The quality of the lumber will not vary much, since most builders use construction-grade wood for joists. But the width of joists is a different matter.

Some builders use 2×8 joists, while others use 2×10s. The distance that the joist has to span has something to do with this decision, but sometimes it is a question of meeting minimum requirements or doing a little extra. Obviously 2×10s are better, not only because they may add strength to the house but also because you may need room in the joist structure to cut holes for plumbing pipes. When you cut a hole through a 2×10, you have a lot more wood left over than when you cut through a 2×8. Some kit owners have said that they had to provide extra bracing in the joist system because the 2×8s weren't wide enough to take the plumbing holes.

SUBFLOOR. Subfloor plywood may not be included in the kit either, but if it is, you should pay attention to thickness. Some kits use ½-inch plywood in the subfloor, while others use ⅝-inch, and still others offer ¾-inch. The thicker the plywood, the more solid the subfloor. If you plan to cover the plywood with a hardwood floor, then this variable is less important, but if you plan to use carpet, you may be able to feel a difference between ½-inch and ¾-inch.

A few kits offer tongue-and-groove plywood. The sides of each piece interlock with the next piece, instead of merely butting together. Tongue-and-groove can add stability and also reduce the amount of air infiltration through the floor.

There are some subfloor systems that don't use plywood at all—frequently, subfloor and floor are one and the same—but they will have to be evaluated independently. The most common system is joists set at 16-inch intervals and covered by sheets of plywood.

STUD WALLS. The majority of house kits use the standard 2×4 studs on 16-inch centers. A few wall systems, particularly post-and-beam systems, use 2×3 studs, which decreases the thickness of the wall and leaves less room for in-wall insulation. It may be that enough outside insulation is provided to make up for this, but this should be considered carefully.

A few kits offer an extra-thick 2×6 stud wall that permits more insulation (although, of course, it diminishes available interior space). This type of wall system is part of a nationwide plan to make houses more energy-efficient.

Some kits, like log cabins and domes, do not use conventional stud walls at all. These kits must be judged on their own terms.

SHEATHING. Some sort of material is customarily sandwiched between the outside of the stud walls and the visible, exterior siding. Sheathing increases the insulation value of the wall and also helps strengthen the house. It can be made of plywood, tar-impregnated particle board, rigid Styrofoam, or a number of other materials. Sheathing is sold in various thicknesses, and when you are dealing with the same material, thicker is always better. The best way to compare two different types of sheathing is to price them at the lumber store. Some companies will skimp on a hidden element like this, while others will offer the best.

SIDING. Kits send anything from basic plywood products to fancy cedar or cypress siding. The arrangement of the siding—vertical or horizontal, board-and-batten or lapped—is not as important an indicator of quality as the type of wood used. Another call to the lumber yard will tell you all you need to know.

CLOSED WALL VS. OPEN WALL. There are two basic wall systems manufactured by kit companies. The first is the closed wall, which arrives at the building site already complete both on the outside and the inside. The closed wall eliminates a couple of carpentry steps. It already includes insulation, and may or may not include wiring and plumbing. Log cabins could be considered a closed-wall system, since the log itself takes the place of exterior siding, sheathing, studs, and interior finish material.

The second, and most common, is the open wall, which may be finished on the outside but is bare to the studs on the inside. The builder puts in the insulation, the wiring, the plumbing pipes, and whatever interior surfacing he chooses.

Each system has its advantages. Closed-wall kits may be quicker to erect. The problem with closed walls is that if they do include plumbing and electric, the units tend to be very heavy and hard to handle. If pipes have been damaged or punctured during handling, they are difficult to reach. And if the closed walls do not include wiring and pipes, then the owner is limited in where else he can run such pipes. He must find enough interior partition walls to hold all the plumbing and wiring, and that is sometimes a challenging task. (In log cabins the wiring is sometimes put behind the baseboards, and the electrical boxes are stuck onto the wall rather than embedded in it.)

Open-wall kits allow more flexibility in the placement of wiring and pipes, because they can be run through all the outside walls as well as through interior partitions. Open walls also permit the builder to see that everything has been done correctly before he boards up the inside.

ROOF SYSTEMS. For house kits that use standard trusses, one of the main variants is the thickness of plywood on the skin of the roof. The thicker the roof, the better. Some kits employ double-roof systems —two layers of plywood with air space in between.

For houses with cathedral ceilings, pay attention to the type of wood used in the decking. Most kits with exposed beams and open ceilings use 2×6 tongue-and-groove decking, although some use 3×6 decking, which is heavier and sturdier. The wood can be relatively inexpensive, like pine, or fancy, like cedar. There are even gradations of quality within one type of wood. Select pine decking has no visible knots and is much more expensive than No. 2 pine decking, which does show knots. You can find out from the kit company which grade of decking is used.

The insulation for an exposed ceiling goes on top of the roof, since there is no attic. We will discuss insulation in the next section on energy. But you might give extra points for exposed ceilings that also have a layer of plywood above the decking, and some air space in between. These sandwiches can add a great deal to the energy efficiency of a house.

SHINGLES. Most kits offer asphalt shingles, with wood shingles as an option. Asphalt shingles are sold by weight. A heavy shingle will carry a longer guarantee than a lighter shingle, and will also cost more. The best ones are guaranteed by the manufacturer for 25 years, while a cheaper shingle may be backed for only 10 to 15 years. Superior shingles usually weigh more than 300 pounds a "square"—the unit of measure for shingles, 100 square feet of finished roof—while mediocre shingles weigh less than 200 pounds.

WINDOWS. Kit companies use so many different kinds that it would be impossible to summarize all the options. A few things can be said. Single glazing (one naked pane of glass) has become obsolete with most companies, since single-glazed windows permit much heat loss. The few companies that still push such windows also offer double glazing as an "extra." Consider double glazing a necessity.

Uninsulated aluminum windows are generally inferior to wooden windows, since wood is a better insulator than aluminum. If your kit offers aluminum windows, find out if the windows include some sort of thermal break between the layers of metal to retard heat loss. Aluminum windows with thermal breaks may compare favorably with wooden windows as energy conservers.

Wooden double-hung windows are popular components in kit houses, and most of them work quite well. Wooden windows can, however, stick or swell if they are not made correctly or if they are not protected from the elements. One of the most successful solutions to this problem is the double-hung window with double glazing and with a vinyl protective covering for the wood on the outside of the house. Leading manufacturers of such windows include Andersen and Pella. They are generally more expensive than other types of manufactured windows, but many buyers report that the extra protection and quality are worth the money.

Energy

Almost every builder and contractor in America now calls his product an "energy saver." Kit companies are no less eager to use the term. Since "energy saver" has no exact definition, it can cover a multitude of cracks. Some kit companies say their houses are energy-efficient because there is double glass in the windows and 3½ inches of insulation in the stud walls. Other houses are called energy savers because they have double roofs, extra solar features, and 6 inches of insulation in the walls. The tribute is bestowed both on companies that comply with minimum standards and on companies that have truly innovative approaches to energy problems.

The most popular measurement for energy efficiency is the well-publicized R-factor, which indicates the effectiveness of any material in resisting the flow of heat through it. You want to know the cumulative R-value for the complete wall package and the complete roof package, and not just the R-value for the insulation. Kit companies put such numbers in their brochures. Northern Homes, for instance, offers a standard R-13 and an R-38 ceiling. Habitat's new Colony Series houses include R-22 walls and R-30 ceilings, with optional R-38 ceilings. The higher the R-factor, the better the insulative barrier, at least to the degree that R-factors are accurate. Recent studies suggest that the R-value of any material can fluctuate wildly depending on the atmospheric conditions around it. When fiberglass insulation takes in 1% moisture, the R-value may diminish by as much as 30% to 40%, according to some researchers. The R-factor is far from a perfect measure, but most building codes and lending institutions are paying a lot of attention to it.

To add to the confusion, building people and banking people do not agree on how much R is a good thing. According to a *Popular Mechanics* survey published in September 1978, the Federal Housing Administration requires R-22 for ceilings (R-19 in areas with less than 8,000 degree-days) and R-13 for walls (R-11 in areas with less than 8,000 degrees days). The Edison Electric Institute pushes R-30 ceilings (R-19 for areas with less than 3,500 degree days) and R-11 for walls. Owens-Corning, the insulation people, suggests a range of between R-13 and R-19 for walls, and R-26 and R-38 for ceilings, depending on what part of the country you live in. For the northern climate zones, the Farmer's Home Administration requires a minimum of R-19 in the walls and from R-30 to R-38 in the ceilings. State, local, and regional building codes all have their own ideas about what R-values are acceptable.

One attempt to resolve these differences has come from the American Society of Heating, Refrigeration, and Electrical Engineers (ASHRAE). ASHRAE has written a model code for energy, called 90-75, and it looks as if many of the state and regional building codes will adopt it. You can check with your local building-code enforcement office to see whether ASHRAE standards apply in your area.

Many kit packages already meet ASHRAE standards. Since kit companies usually have to satisfy more than one state or regional code, they may be quicker to upgrade their insulation levels to comply with new requirements than the more parochial builders.

In any case, R-value is not the last letter in predicting home fuel bills. Heat loss is too complicated a subject to be reduced to a debate over the thickness of blankets. Some kit companies can give you technical information on the total heat loss in their homes, and these calculations are a lot more meaningful than R-factors. Pan Abode Homes, for instance, computes the heat loss (in Btu per hour) from walls, windows, doors, roof, and floors, and comes up with a composite loss of 29,350 Btuh. The only other variable to consider, besides the square footage in the house, is its location. Obviously, a house in a warmer climate will lose less heat than the same house in a colder climate. The Btuh calculations must be adjusted for degree-days, a way of comparing the outside temperature in one part of the country to that in another. Degree-days are a measure of how many degrees the mean daily temperature deviates from the standard of 65° F. If the average low for a certain day is 20° F., and the average high is 60° F., then the mean temperature is 40° F., or 25 degree-days. The sum total of degree-days for a year is the annual heating load.

If you can find other kit companies that keep total-heat-loss numbers, you can compare the energy efficiency of one house directly to that of another. Companies that do such calculations will be able to provide charts for degree-days and show you how to use them. The only catch is that only a small percentage of kit companies do detailed heat studies. You will probably have to rely on some general guidelines, including the following.

Infiltration is one thing you should investigate. Up until recently, it has been assumed that most of the air leaks through cracks around windows and doors. That assumption has sold a lot of caulk. New information suggests that the worst infiltration can happen around electrical boxes, the soleplate (place where the floor and walls come together), kitchen and bathroom exhaust vents, dryer vents, and the chimney or fireplace. The Texas Power and Light Company did an air-leak study in a home of 1,728 square feet and concluded that of the air that escaped, 25% escaped around the soleplate; 20% seeped through electrical outlets embedded in the wall; 13% got out around the duct system; 6% through the kitchen hood vent; 6% around the fireplaces; 3% around the dryer vent; 1% through the bathroom vent; 5% through recessed spotlights; 13% around the windows; and only 4% around doors. The figures will vary from house to house, but the thing to remember is that you can seal your doors and windows with 6-inch gaskets and still have a very drafty building — depending on how the building is designed and constructed.

The leakage around the soleplate and electrical boxes suggests that outer sheathing can be more important than a lot of people think. Some companies use a flimsy sheathing with a low R-value, and then fill the stud-wall cavity with plenty of fiberglass insulation. The fiberglass is effective, as far as it goes, but it cannot reach behind the electrical boxes and it cannot fill the crack between the walls and floor — where infiltration is said to occur. Sheathing does fit between the electrical boxes and the outside of the house, and it may extend down below the floor level to cover the gap around the soleplate. Dow makes a tongue-and-groove Styrofoam sheathing that provides an effective barrier, and some kit companies have adopted it. When evaluating energy factors, you should ask what kind of sheathing is being used, the R-value of the sheathing itself, how it is applied to the studs, and whether it extends below the gap between the walls and floor.

You might also ask how kitchen and bathroom exhaust is handled. Many kit companies do not provide venting devices, so the decision will be yours. Some venting systems require fewer air changes than others, meaning that the furnace does not have to work as hard to replace valuable hot air that has been shipped outside with the odors. Some venting devices are internal, and do not require outside exhaust at all.

Windows are worthy of some attention, especially since many kit companies provide better-insulated windows as an option. Double glazing is mandatory in areas with any sort of winter at all. If you buy windows with single uninsulated panes of glass, make sure you also get storm windows, which accomplish the same thing as double glazing and in some cases do the job better.

A few companies are beginning to offer triple glazing, or a double-paned insulated window covered by a storm window. This arrangement combines the best features of double glazing and storms. It is effective in reducing fuel bills. Timber Kit, designed for energy efficiency, uses triple glazing in its houses.

When you investigate windows, you should ask not only about the glass, but also about the casing and frames. The advantages of double glazing are limited if you have narrow aluminum frames that leak a lot of heat themselves. Unless aluminum windows have some special insulating feature, wood is a superior insulator.

The exterior doors you put into your house kit should also be insulated and well sealed. Timber-Kit offers an air-lock entry, a double-entry system that keeps furnace heat inside the house even as people walk in and out.

So much for the specific items. They all have an incremental effect on fuel bills, but they may not have nearly as much impact as the design of the house and where you put it. We won't get into one of those discussions about southern exposures, northern windbreaks, energy landscaping, and passive technologies, because the information is already out there. The kit company will probably advise you on the best place to put your house on the land anyway.

The effect of house design on energy is another one of those complex topics that reduces to common sense. Kits that are built with high ceilings and large expanses of glass on all sides are going to cost more to heat, per square foot, than kits that are tighter and more compact. Kits with no overhang and a lot of fixed windows are going to create higher air-conditioning bills that kits that are open and breezy. Usually, houses are designed with a specific region in mind, so the fact that kits can be built anywhere creates a need for extra caution. If you are unsure about the energy characteristics of a house kit, find out where the design *originated,* and ask some local builders if it would be suitable for your area. All the theoretical confusion can also be resolved by a couple of visits to house-kit owners. They can tell you exactly what their fuel bills are, where the drafts originate, and where the cold spots can be found.

Kit companies do not customarily provide heating systems, water heaters, or appliances. Intelligent choices can save much more money than a little extra insulation. Again, when you buy a kit, you have the advantage of some extra expertise. Kit people may help you choose the best heater and the most efficient refrigerator. They can help you calculate 15-year fuel savings and weigh those savings against the original purchase price of furnaces, stoves, air conditioners, refrigerators, and other appliances.

Most of the foregoing energy considerations apply to conventional houses and traditional designs. It is also important to mention three unusual designs — solar houses, domes, and log cabins. People who sell kits for these structures all claim that they produce lower fuel bills than regular houses.

SOLAR HOUSES. There are a lot of solar-house plans, but there aren't many solar-house kits. We've been told the reason: It's hard enough for an amateur to nail walls together, let alone take on the plumbing, electrical work, roofing, and solar theory necessary to install a solar collector system. One company, Green Mountain Homes, does sell a solar kit, based on passive technology. Passive technology isn't necessarily a big deal. A window is an example of passive technology. The sun shines through, and you get warm, without having to channel the heat through complicated copper pipes and valves and tanks. The Green Mountain system is a little more elaborate than that. It uses a solar slab, a patented design that the company wants to keep to itself. All we know is that the slab stores heat, and a special shutter can open or close it to the sun, depending on the time of year. The system can be reversed for summer cooling.

We have heard some positive things about Green Mountain. The company reports that "all the houses can be erected by people willing to do the work themselves."

DOMES. The dome manufacturers all claim superior fuel savings. Tate Miller, Cathedralite president, has said that domes use 30% to 50% less energy, because the shape of the dome permits the heated air to travel in a natural path. There are no nooks and crannies where heated air can be trapped and wasted. We have received no information to contradict these assertions.

LOG CABINS. Log-cabin people also claim energy savings, but they have had to get defensive about it. That's because a lot of housing experts believe in R-values, and log cabins generally have poor R-values. Wood is a good enough insulator relative to concrete or rocks, but it takes a thick piece of wood to achieve the same R-value as a conventional stud wall. The ASHRAE handbook gives an R-value of .91 for each inch of maple, oak, or other hardwood, and 1.25 for each inch of fir, pine, or other softwood. The logs in cabins vary in thickness from 5 to about 8 inches, so you can figure a total R-factor of between 5 and 8, as opposed to the 13 to 19 found in stud framing with fiberglass insulation, sheathing, and exterior siding.

Log people fight these numbers with another theory. They say that logs have thermal advantages that go beyond R-values. Logs are solid, for one thing, which means that log walls do not have internal drafts caused by holes, cracks, or improperly installed insulation. Logs are also massive. The log people say that the mass of the logs helps them retain heat better than a thinner standard wall. Once the cabin is toasty, they postulate, it will not lose the heat as quickly as a standard frame house.

We have no definite position on this debate, and again we suggest a few visits as a way of resolving the question on a local level. Many log-kit owners do claim low heating bills and challenge other homeowners to get the same results.

A few of the log companies have tried to combine the best of two worlds with logs on the outside, a wall cavity in the middle, and more logs on the inside. This design permits the addition of insulation in the wall cavity, and provides a place to run plumbing, wiring, and heating systems. Log cabins built in this way can achieve the benefits of log mass, plus a very high cumulative R-value.

Other log companies provide log facing on a conventional stud-wall structure. The logs in these kits are used more like exterior siding than solid structural members.

When either of these unusual designs is offered, we will mention it in the specifications section of this book.

6
A CATALOG OF KITS

A Note About This Section

Specifications, suggested floor plans, and photographs included in this book were provided by house-package manufacturers. Some manufacturers were able to supply photographs only of slightly customized models rather than standard models for which we had specifications. When this occurred, an attempt has been made to indicate discrepancies between standard and optional features.

Space considerations meant that not every item of trim and hardware offered with standard packages could be listed, but major components are covered. All manufacturers, for example, provide construction guides, assembly manuals, and/or blueprints to assist do-it-yourselfers. Most companies stress that floor plans for their models are suggestions only; many offer custom design services for alterations, or individualized house plans.

Site preparation, foundation, and all utilities are the responsibility of the customer unless otherwise noted.

Prices quoted were in effect in late 1981 but were expected to increase in 1982 to reflect fluctuations in the cost of lumber and components. Price-per-square-foot comparisons should be made on the basis of type of package: A partial shell kit will require more finishing expense than a complete package, and some manufacturers include as optional items certain components you may find are standard elsewhere.

Absence of a formal warranty does not necessarily imply that a company does not stand behind its products. Many small firms claim they have no problems in this regard and will gladly rectify any difficulties because their customers are their most effective advertising. Many components such as windows, doors, and shingles carry their own warranties, and these are passed along to the customer.

Wherever possible, we have included information about shipping, unloading, construction time, and any special tools required. We recommend, of course, that you study the manufacturer's literature carefully, compare quality of components and prices, talk with someone who has worked with the product, and consider realistically your budget, your needs, and your skills before you take the plunge into do-it-yourself home construction.

A final note: Neither the authors nor the publishers of this book specifically recommend any of the houses featured here. We present them for your interest and your convenience, and urge you to make your own decision on the basis of careful, considered examination.

Timberpeg Barnhouse

Type of Structure: Two-story post-and-beam shell package

Manufacturer: Timberpeg
Box 1500
Claremont, NH 03734
(603) 542-7762

Materials Provided: Eastern white pine timbers with interlocking mortise-and-tenon joints, pegged with square oak trunnels for frame. Exterior walls resawn pine board-and-batten, pine tongue-and-groove, or cedar clapboard; interior tongue-and-groove pine or Homasote. Roof and wall systems of tongue-and-groove kiln-dried pine, 2″ rigid isocyanurate insulation with both sides foil-faced, strapping. Single ribs, hand-split Western red cedar shakes for roofing. Double-pane insulated windows with screens, skylights, pine doors, patio doors of tempered insulating glass. Hardware and trim.

Exterior Dimensions: 40 × 20′

Living Area: Upper   800 sq. ft.
           Lower   800 sq. ft.
           Total 1,600 sq. ft.

Price: $34,000 FOB Claremont, NH

Price/sq. ft.: $21.25

Warranty: For quantity and workmanship, 30 days

Special Features: Timberpeg also manufactures Cluster Sheds, which are designed to be combined with each other and larger models to provide additional bedrooms, studio space, or other permutations. Particularly good for sloping sites. Other Timberpeg models range up to $47,500 for an imaginative combination of Saltbox and Cluster Shed offering 2,312 sq. ft. of living space.

Energy Features: Design of walls and roof offers increased insulation. System includes layer of tongue-and-groove and layer of 2″ insulation board separated by strapping to create air space. Insulation values said to be R-21 for walls and R-20 for roof. Optional 3″ insulation.

Notes: Timberpeg materials are good quality, details are attractive, and simpler designs such as Saltbox, Gambrel, and Barnhouse are possibilities for do-it-yourselfers. Network of independent sales representatives or consultation with manufacturer suggested.

Price of Information Kit: $10

FIRST FLOOR   40 × 20

SECOND FLOOR

Pan Abode Aristocrat II

Type of Structure: One-story log shell package

Manufacturer: Pan Abode, Inc.
4350 Lake Washington Blvd. N.
Renton, WA 98055
(206) 255-8260

Materials Provided: Wall system of air-dried Western red cedar timbers, 3×6″ or 4×6″ nominal dimensions, single-tongue-and groove with interlocking cross joints (exterior and interior walls). Floor system of 2×10″ hemlock/fir floor joists, plywood subfloor, particle-board underlayment. Roof system of precut gable timbers, #1 Douglas fir beams, 2×6″ hemlock tongue-and-groove decking, 3½″ rigid insulation, plywood sheathing, felt, asphalt shingles (cedar shakes optional). Prehung raised-panel Western red cedar interior and exterior doors. Wood-framed insulated sliding windows with screens. Miscellaneous hardware, interior and exterior trim, caulking, and stain. Complete drawings and 68-page construction manual.

Exterior Dimensions: 79 × 39′

Price: $50,400 (4×6″ wall timbers) FOB Renton, WA

Price/sq. ft.: $24.49

Warranty: One year

Special Features: Options include cedar shakes, skylights, cedar decks, insulation package, triple-glazed windows, and garage. The Aristocrat II is one of Pan Abode's largest single-story models. Two-story models, condominiums, townhouses, and duplexes are also available and are detailed in the information kit. Custom design services are also offered.

Notes: Shipping by flatbed truck, vans, or containers (to Alaska, Hawaii, and overseas). A forklift or boom crane is required for unloading. Company says these homes are "designed for owner assembly." Construction time estimated at an average rate of 100 sq. ft. per day using two workers and ordinary carpentry tools.

Price of Information Kit: $10

Cathedralite Cuesta I

Type of Structure: Panelized shell package for 35′ geodesic dome

Manufacturer: Cathedralite Domes
820 Bay Avenue, Suite 302
Capitola, CA 95062
(800) 538-0782
(408) 462-2210 in CA

Materials Provided: Sixty preassembled triangular space frames consisting of ½″ exterior-grade plywood over kiln-dried 2×4 Douglas fir studs, predrilled for bolts. Riser walls, beveled base plates, bolts, straps, nuts, washers, two skylights, canopies, canopy weather guards and flashings, and four sets of plans also included. Owner supplies interior finish, roofing material, windows, doors, and insulation. Model shown has five openings and three dormers plus 3′ riser wall.

Exterior Dimensions: 35′ diameter, 18′ high

Living Area: 1,286 sq. ft.

Price: $8,930 FOB Medford, OR, shell only

Price/sq. ft.: $6.94

Warranty: One year for materials and workmanship

Special Features: This kit has eliminated much of the time-consuming aspect of dome construction by panelizing the frame and skin in one unit. Domes range from 26′ to 60′ and have been used commercially as well as residentially. Options include extensions, cupola, solarium, garage unit, additional skylights, canopies, and flashings. Company also offers its domes in 2×6 framing to permit extra insulation in colder areas; cost for 35′ dome is $10,270.

Energy Features: Dome shape allows more efficient air circulation than rectangular housing. Cathedralite estimates its dome can be heated or cooled for 30-50% less than rectangular structure of comparable size. Optional line of solar products available.

Notes: Previous building experience not required; Cathedralite says four or five workers can erect a dome shell in eight hours. No special tools needed. Recommends roofing, plumbing, and wiring be installed professionally. Company will compute shipping charges.

Price of Information Kit: $5

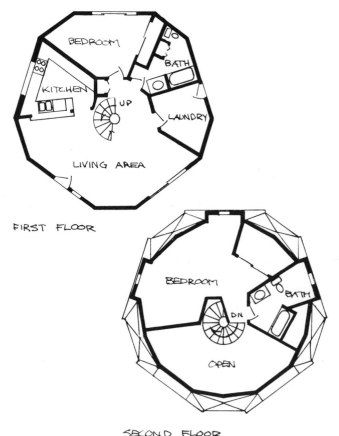

FIRST FLOOR

SECOND FLOOR

Shelter-Kit Unit One

Price of Information Kit: $3

Type of Structure: One-story post-and-beam shell package

Manufacturer: Shelter-Kit Incorporated
22 Mill Street
P.O. Box 1
Tilton, NH 03276

Materials Provided: Floor system of Douglas fir posts, spruce joists, and headers. Studs are spruce and siding is tongue-and-groove pine clapboard. Roofing is exterior-grade plywood sealed with double-coverage roll roofing over Douglas fir, spruce, or yellow pine rafters and joists. Flooring of tongue-and-groove pine. Tempered sliding glass doors; insulating window, screens included. Clerestory of Plexiglas. Angles and brackets, caulking, wood preservative, and miscellaneous hardware and tools also included. Owner supplies the insulation.

Exterior Dimensions: 12×12′

Living Area: 144 sq. ft.

Price: $3,331 FOB Tilton, NH

Price/sq. ft.: $23.13

Warranty: One year for materials and workmanship

Special Features: Options for this model include insulating glass for doors, extra sliding windows and doors, solid-core prehung exterior doors, porch, and deck. Any number of units and options can be joined to create larger homes. Walls are non-load-bearing, which means you can place windows and doors almost at will.

Notes: Truly for the do-it-yourselfer, this unit requires no power tools, can be shipped in one U-Haul truck, and can be built in about four days by two novice builders, according to Shelter-Kit. Shipping bundles weigh about 100 lbs., with precut, drilled, and labeled pieces.

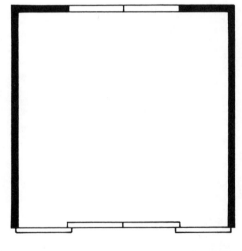

12×12

National Log AL 47

Type of Structure: One-story log shell package

Manufacturer: National Log Construction Company, Inc.
P.O. Box 69
Thompson Falls, MT 59873
(406) 827-3521

Materials Provided: Exterior walls and gables of hollowed, air-dried lodgepole pine, precut for windows and doors, 6" thick and maximum 8' long. Air-Lock system of single tongue-and-groove splines, caulking, and bolts. Rafters of fir or larch. Ridge and purlins solid lathe-turned logs, trusses of top and bottom chords, compression and tension members, steel rods. Window and door frames with half-log trim supplied; owner furnishes window and door units. Porch logs of gables, rafters, walls, posts, railings, and spindles supplied. Miscellaneous hardware and caulking supplied. Owner furnishes roofing, flooring, and subfloor, insulation, finish coat for logs. Garage and interior partitions optional.

Exterior Dimensions: 47×25'

Living Area: 1,280 sq. ft.

Price: $14,546 FOB Thompson Falls, MT. Eligible for FHA, VA, and FmHA financing.

Price/sq. ft.: $11.36; estimated cost of complete house is between $28 and $45 per sq. ft.

Warranty: None

Special Features: Hollow logs permit thorough drying and less cracking later; also handy for electrical wiring, lighter to handle. Air space in center, National Log says, enhances insulation value. National Log offers some 55 models ranging from $4,585 to $50,000, and 80% of its work is custom. Also available are 7" and 8" diameter logs. This firm has been in business for more than 40 years and has another mill in Las Vegas, NM.

Notes: Shipped by truck or railroad car. Recommend about four people to unload. Many do-it-yourselfers build these kits. Estimated construction time two weeks to one month.

Price of Information Kit: $3.50

47 × 25-4

Heritage Washington

Type of Structure: Two-story Cape complete package

Manufacturer: Heritage Homes of New England
456 Southampton Road
P.O. Box 698
Westfield, MA 01086
(413) 568-8614

Materials Provided: Sills and insulating sealer, steel Lally columns and spike-laminated beams for girders; joists and precut bridging. Exterior wall panels 8×12' maximum, prebuilt with sheathing and headers; gables in three sections. Framing lumber Western hemlock. Precut rafters with felt and shingles; siding is lapped and primed hardboard over sheathing paper, many optional sidings. Underlayment provided, oak flooring optional. Precut studs for interior partitions with bearing walls, jacks and interior headers preassembled. All stairs preassembled. Prehung doors, exterior and interior; glued and splined door and window casings. Insulating windows with screens. Mantel, baseboards, bracing, trim, hardware. Optional kitchen and appliance packages.

Exterior Dimensions: 36×24'

Living Area: Upper   624 sq. ft.
Lower   864 sq. ft.
Total   1,488 sq. ft.

Price: $22,402 for materials package unfinished up and shell erection by Heritage builder-dealer. Estimated cost of fully equipped home unfinished up (no site, water, or sewer) is $41,704.

Price/sq. ft.: $15.06 for materials and shell erection: $28.03 for fully equipped version

Warranty: One year for materials; 10-year Home Owners Warranty program sponsored by National Association of Home Builders provided by various builder-dealers.

Special Features: Heritage offers a good deal of preassembly. This model, one of 90 designs, is flexible enough that the upstairs can be left unfinished for future needs. Company has been dealing in New England for more than 25 years.

Notes: Shipping is done in three or four stages, as requested. Says the company, "Heritage, and our builder-dealers, are geared to work along with you and we welcome the opportunity. . . . Any combination of tasks that best fits your abilities, and your time and your pocketbook, can be arranged."

Price of Information Kit: $5

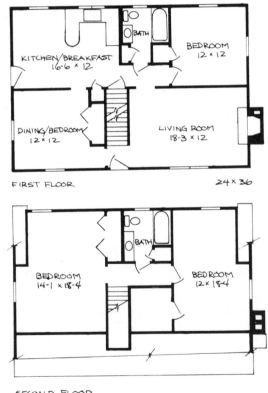

FIRST FLOOR   24 × 36

SECOND FLOOR

Deck House 7123

Type of Structure: Post-and-beam split-level complete shell package

Manufacturer: Deck House, Inc.
930 Main Street
Acton, MA 01720
(617) 259-9450

Materials Provided: Posts are concrete-filled steel columns with top plates for lower level, Douglas fir for upper; beams also fir. Laminated red cedar decking for roof and floor system; sills and plywood subfloor provided. Roof over decking consists of felt, insulating board, more felt, and asphalt shingles. Exterior walls are panelized with V-grooved clear Western red cedar tongue-and-groove over asphalt paper on matrix of studs and plywood sheathing. Fiberglass insulation and vapor barrier between wall and interior sheet rock. Exterior doors solid-core mahogany; windows and sliding glass doors of insulating glass, all in mahogany frames or with mahogany trim and screens. Balcony decking of fir with mahogany millwork. Interior partition studs included, also mahogany stairs and miscellaneous interior and exterior trim and hardware. Model shown has optional carport.

Exterior Dimensions: 28×30'

Living Area: Upper   835 sq. ft.
Lower   746 sq. ft.
Total   1,581 sq. ft.

Price: $25,000 for materials package FOB Acton, MA.

Price/sq. ft.: $15.81 for shell package

Warranty: For quality of materials

Special Features: Deck House offers a quality package for designs ranging from $23,000 to $54,000 (materials only). Passive solar homes are available in the Conservatory Collection; component packages range from $36,000 to $54,000. A national company, its approach is highly systematic and detailed; Deck House works directly with clients through representatives rather than through dealer-builders. Builder Orientation Programs are held for first-time builders and do-it-yourselfers.

Energy Features: This company approaches energy conservation on the basis of return for investment. Roof system is R-21.57; walls R-16.75 with emphasis on reduction of air infiltration through interlocking rather than butt joints. Windows are framed with one solid piece.

Notes: Shipped on tractor-trailer in two loads. About 15% of these packages are sold to people acting as their own general contractors.

Price of Information Kit: $10

FIRST FLOOR

SECOND FLOOR

Boyne Falls Virginian

Type of Structure: One-and-a-half-story preassembled log package

Manufacturer: Boyne Falls Log Homes, Inc.
Boyne Falls, MI 49713
(616) 549-2421

Materials Provided: Air-dried and hand-peeled Northern white cedar and lodgepole pine walls with cedar spline and caulking, prefabricated in post-and-sill method with interior side planed flat (other log styles offered). Walls are shipped 75% preassembled; insulated gable ends preassembled but shipped partially disassembled. Roof system of tongue-and-groove paneling covered by urethane foamboard, furring strips, plywood, asphalt shingles. Floor system of girders, sill plates and sealers, joists, fiberglass insulation, plywood subfloor, felt vapor barrier, tongue-and-groove oak finish flooring. Also partition walls, ceilings, rafters and ties, interior and exterior doors. Windows insulated, with screens, factory preassembled. All hardware and miscellaneous trim included.

Exterior Dimensions: 40×26'

Living Area: Upper   416 sq. ft.
Lower 1,040 sq. ft.
Total   1,456 sq. ft.

Price: $44,130 FOB Boyne Falls, MI

Price/sq. ft.: $30.30

Warranty: 10 Years for exterior wall, if clear-through crack develops.

Special Features: The price seems high, but this is really a complete package, with both floor and roof system and many fine details. All Boyne Falls models available in four exterior finishes—3½" vertical log, 3¼" sill and post, 5½" horizontal log, and 6" lodgepole pine at varying prices. Company established in 1946 and known in commercial building field. National distribution. Northern white cedar is almost maintenance-free.

Energy Features: Minimum of 19% less fuel required to heat or cool than for conventional home, company says. Low thermal conductivity and slow char penetration add extra fire resistance, company says.

Notes: Shipped in two loads, staggered delivery cost is $1.50 per loaded mile. Only basic carpentry tools and experience required. Construction time averages three to four weeks.

Price of Information Kit: $5

FIRST FLOOR   40 × 26

SECOND FLOOR   16 × 26

The Outdoors People Ravel

Type of Structure: Panelized exterior package for 42' geodesic dome

Manufacturer: The Outdoors People
26600 Fallbrook Avenue
Wyoming, MN 55092
(612) 462-1011

Materials Provided: Precut exterior panels, color-coded hubs, 16" Truss Strut framework with attached connectors, 42" riser wall with sheathing, precut 12" thick fiberglass insulation, precut interior sheathing, ventilation cupola, double-glazed windows, precut window jambs, insulated doors, all-weather wood foundation (full basement), first-floor framing, asphalt tabless shingles, roofing paper, flashing, sealant, hardware, tools, and engineer-certified plans. Options include air-lock entry, garage, skylight cupola with five awning windows, foundation insulation, pine or cedar tongue-and-groove interior paneling, triple-glazed windows, cedar shakes. Model shown has optional garage.

Exterior Dimensions: 42' diameter, 21' high

Living Area: Upper   580 sq. ft.
Lower   1,270 sq. ft.
Total   1,850 sq. ft.

Price: $31,283 for exterior package. Also available: shell package, $11,725; complete exterior/interior package, $49,585.

Price/sq. ft.: $16.91 exterior package; $6.34 shell; $26.80 exterior/interior package.

Warranty: One year for materials and workmanship

Special Features: One of the largest and oldest dome companies, The Outdoors People offer a two-day dome-building program in Minnesota, and construction and long-term financing. Their seven residential domes range from 26' to 49' in diameter and can be customized and combined. T.O.P. also provides near-built and full construction services.

Energy Features: The Truss Strut system reduces air infiltration while providing a wall cavity for the ventilation system and 1' of insulation. The Outdoors People can adapt domes for earth sheltering and for active and solar heating systems. This company also offers a unique line of earth-sheltered contemporary houses.

Notes: Dome packages are shipped on 40' flatbed trucks, FOB Minneapolis at $1.85/mile. According to the company, four people can erect the frame in one day and have the exterior finished in one week.

Price of Information Kit: $7.50

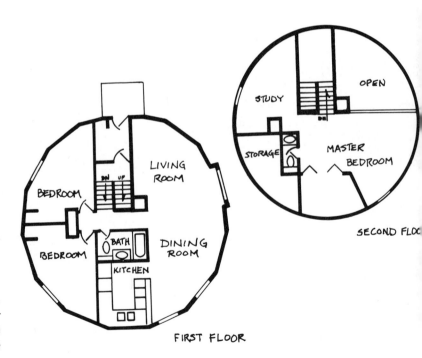

FIRST FLOOR

SECOND FLOOR

American Timber Sunspot 1001

Type of Structure: One-and-a-half-story panelized solar shell package

Manufacturer: American Timber Homes, Inc.
Box 496
Escanaba, MI 49829
(906) 786-4550

Materials Provided: Air-based solar system with collectors, heat exchangers, pumps, blowers, rock storage, water heater and auxiliary heater, and necessary hardware. House package includes 7½"-thick panelized walls with rough-sawn solid tongue-and-groove cedar siding, 2×6 studs, plywood sheathing, ¾" tongue-and-groove interior paneling in choice of 6 woods. Roof system of dimensional-timber single-plane trusses and 2" tongue-and-groove decking. Joists and plywood subfloor supplied for second floor. Framing for interior partitions and paneling. Windows are triple-insulated glass except double-insulated on south wall, all with screens. Exterior doors metal-clad, insulated, and prehung; patio doors wood-framed and insulated. Interior stairs with railings included, also prehung interior doors and closet doors. Garage shown here optional. Owner furnishes roofing and first-floor decking and also a basement for solar equipment. Insulation following American Timber's recommendations furnished by owner, as is interior finish.

Exterior Dimensions: 26×32'

Living Area: Upper   520 sq. ft.
             Lower   832 sq. ft.
             Total 1,352 sq. ft.

Price: $30,259 delivered within 200-mile radius of Escanaba, MI. Add $1.90 per mile beyond 200 miles.

Price/sq. ft.: $23.27

Warranty: One year for materials and workmanship on basic house

Special Features: Collection system based on requirements of house. Large areas of glass concentrated on southern side; northern exposure planned for minimum heat loss. Three other solar models in this series and a variety of options for garages, decks, and interior finishes. Sunspot model also available in larger size in two styles.

Energy Features: Solar collection systems with attention to siting, triple-glazed windows standard except on south wall, open interiors for efficient air circulation. Package includes wood-burning fireplace with heat exchanger and option of conventional furnace or heat pump.

Notes: Mechanical portion of solar system can be installed by competent plumber, electrician, and heating contractor. Do-it-yourselfers can do basic house package if they have carpentry skills. Bulky, readily available items not included in these packages to keep costs down. Can be delivered on one semi to accessible sites in 21 states.

Price of Information Kit: $6, $10 with wood samples

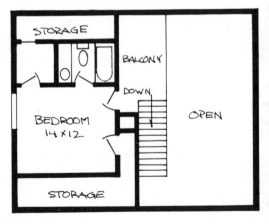

SECOND FLOOR   26 × 32

BEDROOM
9×10

KITCHEN
AND
DINING
15×12

DOWN

BEDROOM
10×11

LIVING ROOM
12×14

UP

RST FLOOR

91

L. C. Andrew Sebec

Type of Structure: One-and-a-half-story prefab log home package

Manufacturer: L.C. Andrew
Cedar Log Homes, Div. K
South Windham, ME 04082
(207) 892-8561

Materials Provided: Two types of kits offered—prefabricated or loose materials. Logs are 3-5″ Northern white cedar, peeled and milled on three sides with matching tongue-and-groove. Logs are beveled top and bottom to shed rain. In prefab homes, logs are fabricated into panels, nailed directly to 3×4 studs with the tongue-and-groove joints precaulked. Insulated steel doors and insulated windows are installed into panels; gable panels and rafters precut. Plywood and felt paper applied to wall and gable end studs under siding. Panels are bored and bolts are furnished for fastening. Single floor and roof sheathing of ½″ plywood. Also included: asphalt shingles, felt paper, and metal drip edge; eave and gable trim; framing materials for interior partitions. For models with lofts or full second floors, materials for stairs, railing, and flooring supplied.

Exterior Dimensions: 32×24′

Living Area: Upper   456 sq. ft.
Lower   768 sq. ft.
Total   1,224 sq. ft.

Price: $19,400 for prefab kit; $16,000 for materials only. Prices FOB South Windham, ME.

Price/sq. ft.: $15.85 for prefab; $13.07 for materials only

Warranty: To dealer, one year for materials and workmanship

Special Features: Northern white cedar does not require preservative, is naturally insect and rot resistant, and surpasses pine and spruce in insulating value. L.C. Andrew's smooth interior walls allow installation of electrical wiring and plumbing into walls as in conventional framed houses. This company has been manufacturing log homes since 1926 and has more than 27 models priced from $6,000 to $38,000. Custom designing is also offered.

Energy Features: Up to R-24 is possible in the walls, depending on the owner-provided insulation. Solar model also available.

Notes: Carpentry experience suggested for do-it-yourselfers. Shipped on flatbed; need five people or equipment for unloading, as well as space at site.

Price of Information Kit: Folder free; complete color portfolio $5

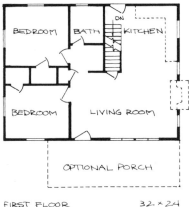

FIRST FLOOR   32 × 24

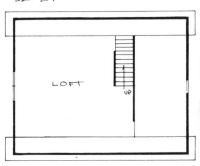

SECOND FLOOR   32 × 24

Miles Jamestown

Type of Structure: Two-story shell package with garage

Manufacturer: Miles Homes
4700 Nathan Lane
P.O. Box 9495
Minneapolis, MN 55440
(612) 553-8300

Materials Provided: Sill plates, laminated beams, joist headers, floor joists, floor bridging, subfloor sheathing, bottom plates, studs, double top plates, interior and exterior walls, stairway, door and window rough framing, ceiling joists, roof braces, rafters, roof and wall sheathing, roof felt and shingles, hardboard exterior siding (aluminum optional), gable and eave trim, soffit and roof louvers, metal-clad insulated doors, double-glazed windows, wall and ceiling insulation, gypsum wallboard, finish floor underlayment, inside trim and hardware for doors and windows, 10' of kitchen cabinets and counter top, miscellaneous hardware and nails. Optional finishing supplies available.

Exterior Dimensions: 25×28' with 20×22' garage

Living Area: Upper 728 sq. ft.
Lower 707 sq. ft.
Total 1,435 sq. ft.

Price: Cash price $33,600

Price/sq. ft.: $23.38

Warranty: Replacement of defective materials

Financing: No down payment. Processing charges about $100; total finance charge for the 24-month contract for this model is $8,020.60 based on an annual rate of 11.9%. The deferred payment price of $41,720.60 is paid in 22 interest-only payments of $364.57 and a 23rd payment of $33,700. The first 22 payments do not reduce buyer's obligation to Miles. Under a second plan, Miles Homes "Build-A-Home Revolving Credit Account and Security Agreement," the amounts and due dates are similar to first schedule. The finance charge is calculated at an annual rate of 11.9%, which is applied to a balance that is equal to what the buyer owes each day of the billing period divided by the number of days in the billing period. In both cases, the final payment is normally paid with long-term financing through a local lender. Miles does not arrange long-term financing.

Special Features: Miles is one of the biggest kit home builders, offering over 50 models nationwide. Its financing plan helps those who might not be able to arrange for credit to finish enough of the dwelling to secure a conventional loan.

Notes: Miles says that if you're skilled with carpentry tools and can follow step-by-step instructions, you can probably handle a good deal of the building yourself. The home is delivered to the site in two shipments when the client is ready for shell and interior-finish steps.

Price of Information Kit: $3

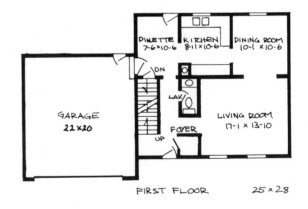

FIRST FLOOR   25 x 28

94

SECOND FLOOR

BATH

BEDROOM
17 × 10-6

DN

BEDROOM
10-2 × 11-1

BEDROOM
10-6 × 14-6

Shelter-Kit Lofthouse 20

Price of Information Kit: $3

Type of Structure: Two-story post-and-beam shell package

Manufacturer: Shelter-Kit Incorporated
22 Mill Street
P.O. Box 1
Tilton, NH 03276
(603) 934-4327

Materials Provided: Floor system of Douglas fir posts, Southern yellow pine trusses, spruce plates. Trusses are preassembled. Roofing of spruce rafters and ridgepole with plywood and asphalt shingles. Walls are rough-sawn reverse board-and-batten plywood siding. Flooring for first and second stories is tongue-and-groove plywood. Sliding tempered glass door and windows. Miscellaneous trim and hardware included. Owner supplies insulation. Model shown is the smaller 16×16′ structure.

Exterior Dimensions: 24×20′

Living Area: Upper 480 sq. ft.
Lower 480 sq. ft.
Total 960 sq. ft.

Price: $8,913 at factory: transportation $1.50 per mile one-way, $100 minimum

Price/sq. ft.: $9.28

Special Features: Incremental modules of 8′ can be added to front or rear of basic structure. Deck, porch, and lean-to options available. Insulating glass for doors and windows also available. This is a new product for Shelter-Kit, which also offers the Unit One cabin featured in this book.

Notes: Designed for assembly by inexperienced builders. Shelter-Kit says this Lofthouse can be assembled in about two weeks by two people. Packaged in easy-to-carry bundles.

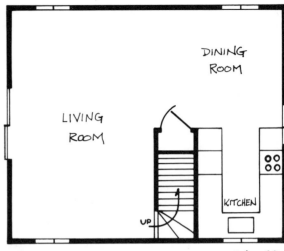

FIRST FLOOR 24 × 20

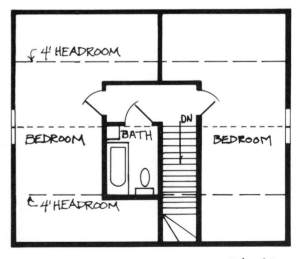

SECOND FLOOR 24 × 20

Northeastern Log Bedford

Type of Structure: Two-story log saltbox package

Manufacturer: Northeastern Log Homes, Inc.
Groton, VT 05046
(802) 584-3336

Materials Provided: Plywood subfloor; spruce joists, girders, headers, and sills. Walls and gables precut and numbered tongue-and-groove Eastern white pine, 6"×8"×16' maximum. Sealing system of splines, PVC foam gasket, caulking, and 10" spikes. Logs treated with a non-toxic preservative. Roof system of pine V-groove boards, spruce strapping, layer of plywood, felt paper, asphalt shingles. Windows are double-hung insulating glass. Exterior and interior doors, interior partitions, and pine V-groove sheathing for interior walls supplied, as are V-groove pine boards for ceilings. Miscellaneous hardware and trim included.

Exterior Dimensions: 37×27'

Living Area: Upper   624 sq. ft.
Lower   936 sq. ft.
Total 1,560 sq. ft.

Price: $27,825 plus shipping

Price/sq. ft.: $17.84

Warranty: None

Special Features: Factory direct sales. No charge for custom design service. Brand-name windows and doors. Double roof system (owner provides insulation) for reduced heat loss. More complete package than most companies offer.

Notes: Packages shipped on flatbed trailers and unloaded for customer (or may be piggybacked by rail); site must be accessible. Sales offices also in Kenduskeag, ME, and Louisville, KY.

Price of Information Kit: Free folder; planning kit $6

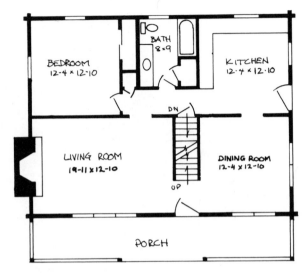

FIRST FLOOR   36×26

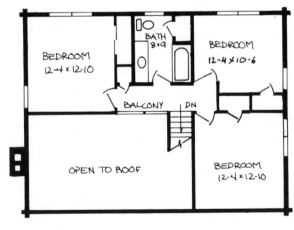

SECOND FLOOR

Northwest Mansard

Type of Structure: Two-story panelized complete shell package

Manufacturer: Northwest Building Systems, Inc.
11702 24th Avenue East
Tacoma, WA 98445
(800) 426-1236
(206) 285-6054 in WA

Materials Provided: Floor system of kiln-dried joists, beams, plywood subflooring, all precut. Exterior walls of framing, building paper, and fir siding (cedar, redwood, and pine optional) in panels of 4', 8' or 12' lengths. Precut roof system of smooth or rough-sawn ridge beams, rafters, and plywood sheathing with felt and shingles. Interior partition framing with trimmers and headers all preassembled, also plumbing wall framing. Precut stairways, railings, fir sundecks. Windows and patio doors are Thermopane, with screens; doors are prehung, solid-core mahogany exterior only. Package includes all roof, window, and miscellaneous trim and hardware.

Exterior Dimensions: 20×30'

Living Area: Upper   480 sq. ft.
Lower   600 sq. ft.
Total   1,080 sq. ft.

Price: $14,949 FOB Seattle, WA

Price/sq. ft.: $13.84

Warranty: One year for materials and workmanship

Special Features: Many options for this special panelized package, including cedar siding, interior walls, and decking; special stud walls for maximum insulation; wood windows and patio doors; shakes for roof. Northwest Building Systems is highly flexible and willing to put together any package. Many models and styles, including a line of modular round Paneloc buildings and round-log wall systems.

Notes: Shipping is by flatbed, about $1 per mile. Precutting and panelizing "tremendously simplifies erection by the inexperienced builder," says Northwest Building Systems. The Mansard is "an amateur's delight for erection" because no crane or mechanical equipment is required. "One man and a helper can easily handle the job," says the company.

Price of Information Kit: $2

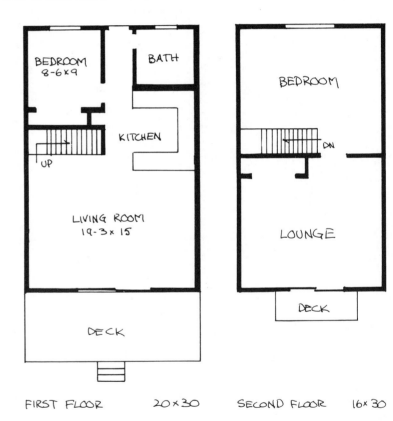

FIRST FLOOR   20×30   SECOND FLOOR   16×30

Model-Log Ruby Gulch

Type of Structure: One-story log shell package

Manufacturer: Model-Log Homes
Lumber Enterprises, Inc.
75777 Gallatin Road
Bozeman, MT 59715
(406) 763-4411

Materials Provided: Round, triple tongue-and-groove logs of lodgepole pine or Douglas fir in 6″ to 10″ diameters. Optional hand-hewn trusses available for open-beamed ceilings. Qualified supervisor comes to construction site for 2 to 3 days to oversee log-wall erection.

Exterior Dimensions: 46×33′

Living Area: 1,248 sq. ft.

Price: 6″ logs, $9,825; 7″ logs, $12,137; 8″ logs, $14,449; 9″ logs, $15,289; 10″ logs, $16,130

Price/sq. ft.: 6″ logs, $7.87; 7″ logs, $9.26; 8″ logs, $11.58; 9″ logs, $12.25; 10″ logs, $12.92

Warranty: Lifetime warranty on logs

Special Features: A deep check-groove is cut in each log to ensure even drying, a special technique that reduces cracking and warping. Eleven standard plans are offered; all can be customized or buyer's plans can be used. Most Model-Log homes are custom designed.

Energy Features: Solid log wall has been determined to be six times as efficient as brick and fifteen times as efficient as concrete in thermal conductivity. A Model-Log home has also been shown to cost 20% less to heat than a comparably sized conventional wood-framed residence. These log homes can also withstand heavy snow loads.

Notes: Freight is estimated at approximately $2 per mile. An experienced crew can assemble log walls in one day.

Price of Information Kit: $5

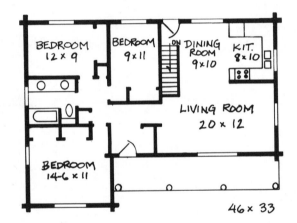

Pacific Frontier Comptche

Type of Structure: One-and-a-half-story post-and-beam shell

Manufacturer: Pacific Frontier Homes, Inc.
17975 North Highway 1
P.O. Box 1247
Fort Bragg, CA 95437
(707) 964-0204

Materials Provided: Post-and-beam framing of 4″ redwood posts and 4″ Douglas fir rafters; redwood tongue-and-groove lumber for exterior and interior walls; 2×6 pine roof deck; 1″ foil-faced urethane insulation for walls and roof; finish roof; wood exterior doors; dual-glazed aluminum sash and patio doors; porch deck and joists; interior partition walls; all special connecting hardware. Not included: floor joists and subfloor, wiring, plumbing, exterior steps and railing.

Exterior Dimensions: 52×44′

Living Area: Loft   400
       Lower 1,424
       Total   1,824

Price: $34,599 FOB Fort Bragg, CA

Price/sq. ft.: $18.97

Warranty: Any unsuitable materials shipped will be replaced. When Pacific Frontier does the on-site construction (in California only), by state law they are responsible for adjustment of any defects for 12 months.

Special Features: The 26 Frontier series homes combine exposed post-and-beam framing with redwood siding inside and out. Also offered is the Cottage series, which uses conventional 2×4 framing, trussed roof, and plywood siding.

Energy Features: The continuous insulation used greatly reduces air infiltration. The standard package satisfies California Energy Commission requirements for all areas with up to 5,000 degree days. Pacific Frontier can provide any required R value in the roof and walls, and will maximize the passive solar potential on a given site through use of such features as glazing and attached greenhouses.

Notes: Kits are shipped by truck and trailer in the West, by rail to the Midwest and East, and by container to off-shore destinations. According to Pacific Frontier, 60% of their clients participate in building their homes.

Price of Information Kit: $3

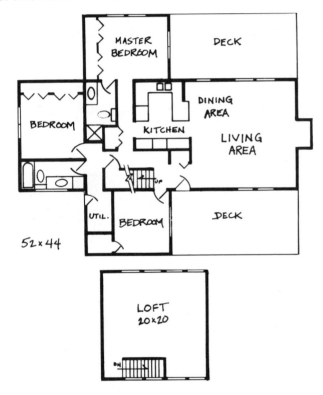

New England Log Barrington

Type of Structure: One-and-a-half story log shell package

Manufacturer: New England Log Homes, Inc.
2301 State Street
P.O. Box 5056
Hamden, CT 06518
(203) 562-9981

Materials Provided: Precut hand-peeled red or white pine logs for walls, second-floor joists, and roof rafters. Wall logs are 8-10'' diameter with maximum length of 12'. Mortise-and-tenon at corners; wall logs tongue and grooved. Sealed with 10'' spikes and gasket material. Pine tongue-and-groove second-floor decking and roofing, heavy-duty asphalt shingles, felt, caulking, and urethane insulation for roof. Owner provides interior finishing materials. Options include full second-floor package, insulated glass, storm windows and doors, dormers.

Exterior Dimensions: 32×52'

Living Area: Upper 704 sq. ft.
Lower 1056 sq. ft.
Total 1760 sq. ft.

Price: $21,190 FOB Great Barrington, MA; Lawrenceville, VA; and Houston, MO.

Price/sq. ft.: $12.04

Warranty: None

Special Features: Wide range of models from $8,000 to $33,000. Four hours of technical assistance given by dealer after delivery and sorting of logs. Extra logs are supplied.

Notes: Site must have room for two 65' trailers to unload side by side. At least three people plus dealer must be present for unloading, and dunnage must be supplied. Tools required include level, 16' and 50' tapes, chalk line, 8-lb. sledge, ''come-along'' and chain, caulking and staple guns, wire brush, Skilsaw, chain saw, hammers. NELHI cautions that builder should have some carpentry experience or be handy with tools.

Price of Information Kit: $5

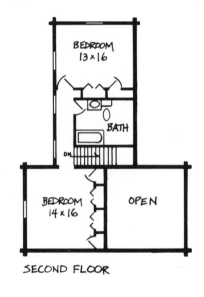

SECOND FLOOR

FIRST FLOOR   32×52

Hainesway Linnea

Type of Structure: One-story log shell package with interior partitions

Manufacturer: Hainesway II Log Homes
Sylvan Products, Inc.
4729 State Highway No. 3, S.W.
Port Orchard, WA 98366
(206) 674-2511

Materials Provided: 6½" diameter logs with attached keys, numbered and ready to assemble; four gable walls; door and window frames specially milled from 2×8" timbers. Interior log partitions also included. All other materials are normally obtained from local building supply companies.

Exterior Dimensions: 46×41'

Living Area: 1,503 sq. ft.

Price: $19,515 FOB Port Orchard, WA

Price/sq. ft.: $12.98; Hainesway says finish price/sq. ft. including labor is approximately $35.

Warranty: None, but the company will correct or replace defective materials promptly.

Special Features: Over a span of 30 years Hainesway has perfected its system of Dual-Key lay-up, mortise and tenon joints for settlement, and interlocking corners. The Dual-Key system uses two recesses on each side of a log that receive keys that join the log with the logs below and above it in the log wall. These keys and a 3" cut in each log prevent warping and twisting while providing a tight fit that can only get tighter with age. Nine models are offered and a drafting service is available. Hainesway's specialty is custom cutting.

Energy Features: The solid log walls have an R value of 22. Because they store heat in winter, the logs ensure a more even interior temperature day and night. Logs also help to cool the house in summer.

Notes: Hainesway points out that the "lower amount of manufacturing energy required to prepare the logs and the simplified construction methods are honest approaches to reducing home building costs." Step-by-step instructions with sketches and a coded parts list provide guidelines for the do-it-yourselfer or contractor. Long-distance shipping is by common carrier; at least two people are required for unloading. Logs can be fully erected in less than a week.

Price of Information Kit: $2

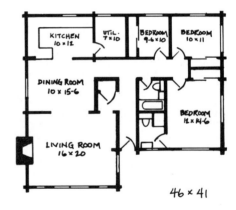

Wickes Lake Haus

Type of Structure: Two-story preassembled shell package

Manufacturer: Wickes Components
Division of The Wickes Corporation
4461 Tilly Mill Road
Doraville, GA 30360
(800) 241-7318
(404) 452-8130 in GA

Materials Provided: Floor includes beams, sill plates, joists, plywood subfloor, stairs, and hardware. Walls are preassembled sections of varying lengths with framing (2×4 kiln-dried lumber) for prehung windows and doors. Gable-end panels, bathroom plumbing walls, and choice of siding included. Roof consists of all trusses and gable ends, plywood, felt, asphalt shingles, overhang and soffit assemblies, hardware.

Exterior Dimensions: 24×30'

Living Area: Upper 277 sq. ft.
Lower 720 sq. ft.
Total 997 sq. ft.

Price: $9,100 for shell; $18,300 for package with interior finish materials plus plumbing, heating, and electric packages, delivered in Michigan

Price/sq. ft.: $9.13 for shell only; $18.36 for interior and utility packages

Warranty: None

Special Features: The basic shell can be expanded in 2' increments at time of construction. Wide range of models and option packages (insulation, insulated windows, appliances, trim). Wickes has dealers in 21 states; offers own financing plan.

Notes: Delivery is in two loads, one on 55' trailer. Wickes offers 26 floor plans in conventional and vacation housing in shell-only price range from about $7,100 to $18,350. "With help from friends or relatives, the Lake Haus can be under cover in a very short time," says Wickes.

Price of Information Kit: Free

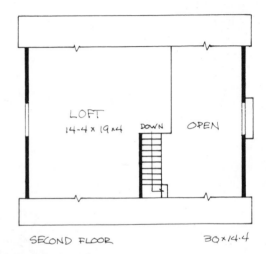

SECOND FLOOR   30×14.4

FIRST FLOOR   30×24

Habitat Passive Solar

Type of Structure: Two-story post-and-beam solar shell package

Manufacturer: Habitat
123 Elm Street
South Deerfield, MA 01373
(413) 665-4006

Materials Provided: Complete pre-engineered shell. Framework is 6×8" Western fir. Walls are preassembled using 2×4s 16" on center with 1" styrofoam sheathing and prestained siding applied on-site (owner supplies fiberglass sidewall insulation). Floor and roof systems are exposed 2" tongue-and-groove planking. Roof insulation is 3" urethane/nailbase; 290# shingles used. Exterior rake, fascia, and trim are 1" rough-sawn cedar. Insulated wood patio doors and windows, operable direct-gain solar roof glazing system.

Exterior Dimensions: 36×28'

Living Area: Upper   624 sq. ft.
Lower   912 sq. ft.
Total   1,536 sq. ft.

Price: $26,098 FOB South Deerfield, MA, shipping about $1 per mile.

Price/sq. ft.: $16.99 ($16.48 with optional third-floor loft) for complete shell package with solar glazing

Warranty: Manufacturers' warranties passed on to kit buyer

Special Features: Optional third-floor loft for additional 336 sq. ft. Exposed post-and-beam framing in a passive solar design.

Energy Features: Direct-gain solar roof, innovative insulation system, and entry air locks add up to substantial energy savings. Solar greenhouse optional. This model was an award winner in the Passive Solar Design Competition sponsored by the U.S. Department of Housing and Urban Development in 1978 in cooperation with the U.S. Department of Energy and the Solar Energy Research Institute.

Notes: Kit can be built by contractor or owner.

Price of Information Kit: $4

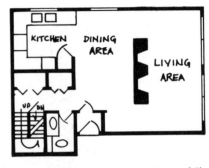

FIRST FLOOR   36 × 28

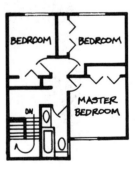

SECOND FLOOR

112

American Timber Country Squire 301

Type of Structure: One-story panelized shell package

Manufacturer: American Timber Homes, Inc.
Escanaba, MI 49829
(906) 786-4550

Materials Provided: Exterior walls are rough-sawn kiln-dried and treated Northern white cedar siding, building felt, sheathing, and 2×6 framing assembled into 8', 9', or 10' panels. Owner supplies insulation, flooring, and roofing. Roof support of hand-hewn fir trusses supported by columns built into wall panels, spanned by 2″ tongue-and-groove decking. Trusses and gable ends assembled. Interior partitions include studs and trim. Windows are insulated. All doors are prehung and have birch interior, metal-clad and insulated exterior. Covered porches complete with overhang, railing, steps, and decking of treated cedar. Hardware and trim included. Solid ¾″ tongue-and-groove paneling available in cedar, aspen, cherry, ash, balm of Gilead, and naturally aged barnwood.

Exterior Dimensions: 40×26'

Living Area: 960 sq. ft.

Price: $27,896

Price/sq. ft.: $26.82

Warranty: One year for materials and workmanship

Special Features: Good-quality optional interior tongue-and-groove paneling and sun decks as well as dimensional timbers instead of log trusses. Although the owner supplies several major components, the basic American Timber package is a strong and durable one. Northern white cedar is low-maintenance. The trusses used in these homes are balsam fir, hand-peeled, seasoned, and hand-hewn with machined steel rings set into the wood for connections. Use of trusses allows free span for large areas and flexibility of interior design. Model shown is one of least expensive—range goes to $90,000-plus.

Energy Features: Because American Timber also offers a line of solar-home packages, a number of details in these houses are heavy-duty with conservation in mind.

Notes: Simpler models can be built by do-it-yourselfers with experience. American Timber ships on its own semis; this model requires only one load. Delivered prices will be quoted.

Price of Information Kit: $6, $10 with wood samples

David Howard's Mulligan Frame

Type of Structure: Braced red oak post-and-beam frame for two-and-a-half story colonial

Manufacturer: David Howard, Inc.
P.O. Box 295
Alstead, NH 03602
(603) 835-2213

Materials Provided: Red oak post-to-truss-to-beam frame, precut and erected by David Howard's crew. Frame is about $11 per square foot in New England, oiled and either planed or gouged for handhewn finish. Frame exterior is covered with rigid foam insulation (3½" for wall, 5½" for roof) and chicken wire for plaster lathe. Owner supplies foundation, floor decking, roofing, exterior and interior finish. Howard recommends plaster for exposed-beam interior and clapboards or cedar shingles for exterior.

Exterior Dimensions: 42×30'

Living Area: Third floor 567 sq. ft.
Second floor 1,001 sq. ft.
First floor 1,260 sq. ft.
Total 2,828 sq. ft.

Price: $30,800 in New England, erected frame. Price includes shipping and crew costs in New England.

Price/sq. ft.: $10.89

Warranty: None

Special Features: David Howard's houses are usually custom-designed, ranging from traditional New England models to contemporary styles (about 70 models). He will supply at extra cost details such as fanlights, Dutch doors with walnut hardware and carvings, cabinets, stairways, furniture, and matched pine flooring. Price includes two site visits, all plans and instructions, and a supervision service for the remainder of the project.

Notes: Owner responsible for unloading, forklift or crane necessary; site must be accessible to crane used in construction. Frame erection usually takes three to four days. Instruction book teaches "the art of carpentry, plastering, finishing, plumbing and wiring. Cabinets, doors, stairs and windows are all ready to install and finish. Don't be afraid—you can do it yourself."

Price of Information Kit: $6 for 116-page catalog with more than 70 designs

SECOND FLOOR

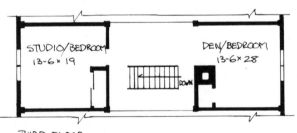

THIRD FLOOR

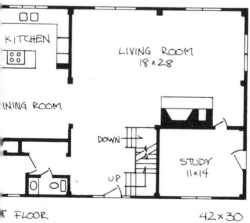

KITCHEN

LIVING ROOM
18 x 28

INING ROOM

DOWN

UP

STUDY
11 x 14

FLOOR   42 x 30

Wilderness Log Homesteader

Type of Structure: One-and-a-half-story log shell package

Manufacturer: Wilderness Log Homes
R.R. 2
Plymouth, WI 53073
(800) 558-5881
(800) 852-5828 in WI

Materials Provided: "Insul-Log" package uses 10"-diameter, treated, quality pine logs cut in half for use with conventional framing as interior wall, exterior wall, or both. Package includes hand-peeled log rafters; beams; rails and posts; interior partitions; knotty-pine paneling for ceiling; choice of cathedral ceiling with double-roof system or exposed-beam conventional ceiling with vapor barrier, sheathing, felt, and asphalt shingles; exterior doors and double-glazed windows. Insulation supplied is 12" fiberglass batts for roof and 6" for walls. Package includes all materials and hardware for finished shell. Model shown has custom porches.

Exterior Dimensions: 32×28'

Living Area: Upper 448 sq. ft.
Lower 896 sq. ft.
Total 1,344 sq. ft.

Price: $22,262

Price/sq. ft.: $16.56

Warranty: Ten years for materials and workmanship

Special Features: This package provides for a full 6" of insulation in the wall cavity and also offers flexibility of finish and ease of wiring. Wilderness also offers full-log packages for conventional log construction. Options include cedar logs, loft or second-story package, cedar shakes, porches, knotty-pine tongue-and-groove interior. This appears to be a good, complete shell package with an unusually good warranty.

Notes: Shipping nationwide on 40' flatbed included in price; one truck required for this model. At least two people recommended for unloading. Average construction time is 10 days to two weeks, according to Wilderness.

Price of Information Kit: $5

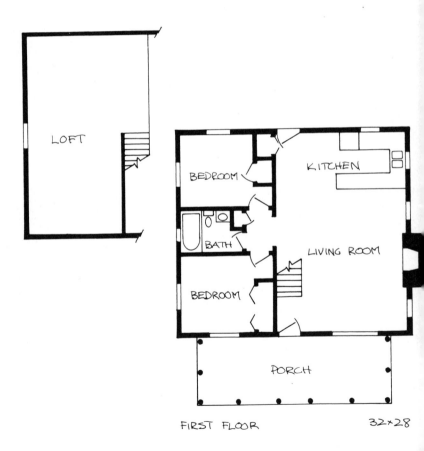

FIRST FLOOR   32×28

Green Mountain N-38

Type of Structure: Two-story panelized solar shell package

Manufacturer: Green Mountain Homes
Royalton, VT 05068
(802) 763-8384

Materials Provided: Customer installs gravel-and-concrete solar slab. Subflooring is plywood over kiln-dried 2×10s. Prefabricated wall panels 8' wide consist of plywood over studs, with tongue-and-groove Styrofoam and rough-sawn shiplapped pine siding for an R-19 wall. Insulation is placed both in stud cavity and between sheathing and exterior siding. Roof is factory-precut plywood sheathing over rafters, vented for moisture control. Owner supplies insulation for minimum R-32, as well as roof covering. Dual-glazed windows with screens, dual-glazed weathertight sliding glass doors, insulated exterior door. Airlock entrance with two doors. Prestained pine timbers provided for second-floor support. Optional thermal shutters raise R-factor of windows or sliding doors from 1.81 to 11.23.

Exterior Dimensions: 39×17'

Living Area: Upper   608 sq. ft.
Lower   656 sq. ft.
Total   1,264 sq. ft.

Price: Approximately $18,290 for kit; estimated total house cost (excluding site work, septic, well, etc.) about $36,800 for do-it-yourselfers and $45,000 if contractor-built.

Price/sq. ft.: $14.47 for kit; $29.11 for do-it-yourself total house; $35.60 for contractor-built house

Warranty: One year

Special Features: Good-quality materials and a lot of thought have gone into this model, from the 4'-deep slab to the optional wood-burning stove and heat-return fireplace backed up by a furnace. All Green Mountain Homes are expandable and attractive. Price range on a do-it-yourself basis is from about $18,000 to $68,000.

Energy Features: Green Mountain's N-38 home cost $249 to heat in the severe Vermont winter of 1976-77. (House can be heated with two cords of wood per heating season in Vermont, less in more temperate climates.) Entire house is a solar collection and storage unit, with multilayered roof and walls, east, west, and south-facing windows, air-lock entrance, and the solar slab designed to hold and distribute warm air in winter and cool air in summer.

Notes: "Can be erected by people willing to do the work themselves," says the company. Shipped on flatbeds, no special unloading equipment needed but four or five people recommended. No special tools.

Price of Information Kit: $4.50

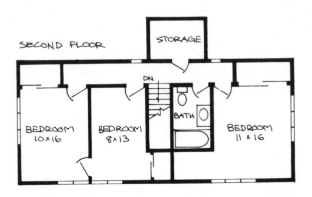

Polydome 1050

Type of Structure: Panelized shell package for 32' geodesic dome

Manufacturer: Polydome, Inc.
3020 North Park Way
San Diego, CA 92104
(714) 574-1400

Materials Provided: Three trapezoids and 24 triangular panels made of exterior-grade ½" plywood over struts of 2×4, 2×6, and 2×8 Douglas fir, plus nuts, bolts, and miscellaneous hardware. Anchor bolts not included. Trapezoids weigh 250 lbs., triangles 115 lbs. Model shown here has three entry openings and 3' riser wall.

Exterior Dimensions: 32' diameter, 18' high

Living Area: Upper   427 sq. ft.
Lower   940 sq. ft.
Total   1,367 sq. ft.

Price: $5,445 for basic shell kit FOB El Cajon, CA

Price/sq. ft.: $3.98; estimated finished cost $18 for do-it-yourselfers

Warranty: One year on workmanship and materials

Special Features: Polydome offers 26', 32', and 40' domes and many options, including skylights, precut tongue-and-groove paneling, heavy-duty snow-load version, completion kit, carport, dormers, eyebrow (pictured), insulation already installed in trapezoids and triangles for R-19 rating, and cabinets.

Energy Features: Dome configuration permits savings, says Polydome, of 35-50% on heating and cooling costs. Sturdier snow-load version $5,995.

Notes: Wrench needed to connect predrilled panels. Wood or pipe scaffolding can be used, although a crane speeds erection time considerably and can usually be rented by the hour. "No previous ex-perience or skill is needed" for Polydome erection, says company, "but assistance from an authorized Polydome assembler is available." Openings designed in and require no extra cutting. Can be shipped in 12' open trailer or truck.

Price of Information Kit: $5

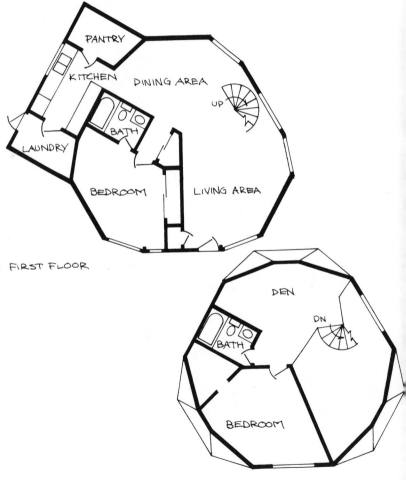

FIRST FLOOR

SECOND FLOOR

Ward Cabin Hawthorne

Type of Structure: One-and-a-half-story complete log package

Manufacturer: Ward Cabin Company
P.O. Box 72
Houlton, ME 04730
(800) 341-1566
(207) 532-6531 in ME

Materials Provided: Precut 5″ Northern white cedar logs, tongue and grooved, flat inside face with V-jointed edges, hand-peeled exterior face, beaded for caulking; 8″ spikes, caulking, and foam gaskets. Sills, floor joists, headers, bridging, preassembled spruce girders; 5/8″ plywood subfloor and ½″ underlayment. Basement stairs and rails; rustic loft or oak loft stairs and rails; interior partition framing of 2×4 spruce with ¾″ pine or cedar paneling. Six-panel colonial interior doors with cedar trim; prehung energy-saving exterior doors with cedar trim; double-glazed windows, precut frames, cedar trim, flashing, screens, vinyl grilles, and shutters. Ceilings use precut 4×8 beams with 2″ tongue-and-groove loft floor over beams. All support posts and loft railings with necessary hardware. Precut roof beams are round spruce purlins. Double-roof system of ¾″ tongue-and-groove pine, 2×3 strapping, 5/8″ plywood. Pine eave trim; drip edge, valley flashing, felt paper, venting systems, and asphalt shingles. Less expensive truss roof available; roof insulation optional. Complete approved building plans and 110-page construction manual also provided with kit.

Exterior Dimensions: 32×24′ with 32×8′ porch

Living Area: Loft   351 sq. ft.
First floor 768 sq. ft.
Total   1,119 sq. ft.

Price: $31,350 FOB Houlton, ME

Price/sq. ft.: $28.02

Warranty: 30 days after delivery for defects, workmanship, design, and quality

Special Features: Ward specializes in custom designing at no extra charge and will quote on a package for your plan at no charge. Over 50 years of experience make this company the oldest log home manufacturer in the industry. Ward has a national dealer network to serve most customers and also sells direct.

Notes: Usually shipped by closed 40′ trailers for up to 1,000 miles at $1.50 per mile; rail may be used for longer distances. Customer hand unloads, four to six people recommended. No special tools required for construction; however, a crane is recommended for installation of roof beams.

Price of Information Kit: Free folder; $6 for 72-page, color plan catalog

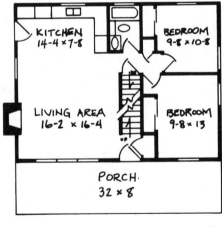

FIRST FLOOR   32 × 24

STORAGE

OPEN

DN

BEDROOM
12 × 11

STORAGE

SECOND FLOOR

Green Mountain N-18

Type of Structure: Two-story panelized shell package

Manufacturer: Green Mountain Homes
Royalton, VT 05068
(802) 763-8384

Materials Provided: Floor system of plywood over kiln-dried 2×10s. Prefabricated wall panels 8' wide consist of plywood over studs, with tongue-and-groove Styrofoam and rough-sawn shiplapped pine siding. Insulation is placed both in stud cavity and between sheathing and siding. Roof is factory-precut plywood sheathing over rafters, vented for moisture control. Owner supplies insulation and roof covering. Dual-glazed windows with screens, dual-glazed weathertight sliding glass doors, insulated exterior door. Prestained pine timbers for second-floor support. Photo shows north side of N-18; south side same as end section of N-38 model, also featured in this book.

Exterior Dimensions: 18×16'

Living Area: Upper 288 sq. ft.
Lower 336 sq. ft.
Total 624 sq. ft.

Price: Approximately $9,350 for kit; estimated total house cost (excluding site work, septic, well, etc.) about $18,000 for do-it-yourselfers and $25,000 if contractor-built.

Price/sq. ft.: $14.98 for kit; $28.85 for do-it-yourself total house; $40.06 for contractor-built house

Warranty: One year

Special Features: This model is designed as a starter house, one you can add to at a later date simply by erecting another shell and cutting openings on each floor. Like all Green Mountain models, it's highly flexible and well designed. This company offers 13 designs, one of which is the N-38 solar model featured in this book.

Energy Features: Green Mountain offers a number of energy-saving ideas such as thermal shutters, multilayered roof and walls, and heat-return fireplaces. Houses are designed with no northern windows but plenty of glass on east, west and south sides.

Notes: This design, says Green Mountain, is so simple that it can be built by an amateur in just a few days, with a little help from friends. Shipped on flatbeds, no special unloading equipment or tools needed.

Price of Information Kit: $4.50

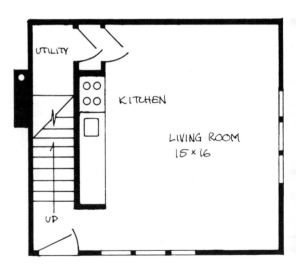

FIRST FLOOR

BEDROOM
6-6 × 11-6

BEDROOM
9×11

UP

BATH

SECOND FLOOR

Bow House Half Cape

Type of Structure: Two-story Cape with ell shell and detail package

Manufacturer: Bow House, Inc.
Randall Road
Bolton, MA 01740
(617) 779-6464

Materials Provided: Bow House's philosophy is to reduce transportation costs by supplying only the items necessary for the character of the house. These include 8-ply laminated Southern yellow pine roof rafters treated and precut, gable rake boards, beehive-design attic ventilator, white cedar shingles, red cedar clapboards, pine trim for exterior and interior, prehung exterior doors, Boston-style windows, front shutters, 12" or wider pine floors, prehung interior doors, stairs, fireplace mantel and accessories, masonry form for chimney cap, base molding, parson's cupboard door, screens, and special trim and hardware. Owner furnishes items such as insulation, floor support, sheathing, studding, fireplace materials, and other items cheaper locally.

Exterior Dimensions: 24×28', ell 20×14'

Living Area: Upper   668 sq. ft.
Lower   952 sq. ft.
Total   1,620 sq. ft.

Price: Package only, $19,500 FOB Bolton, MA. Bow House says finished price (allows for site development, appliances, kitchen cabinets, and light fixtures) will range between $75,000 and $85,000 depending on local prices.

Price/sq. ft.: For package, $12.03; for completed house, from ground up, $46.30-$52.50

Warranty: None

Special Features: Distinctive roof design allows increased upstairs space; in addition to excellent strength characteristics, the roof sheds snow easily. Attention to detail both inside and out makes this an appealing choice. A wide range of handmade or specially made options to complete the interior, including hearth bricks, hand-forged hardware, light fixtures, and wooden storm windows. Other models are full Cape, three-quarter, and quarter.

Notes: Because Bow House includes only those items not available locally, shipping can be done on one truck to reduce transportation costs. Company has shipped throughout the country. While a builder is recommended, many interior finish items can be done by do-it-yourselfers.

Price of Information Kit: $4

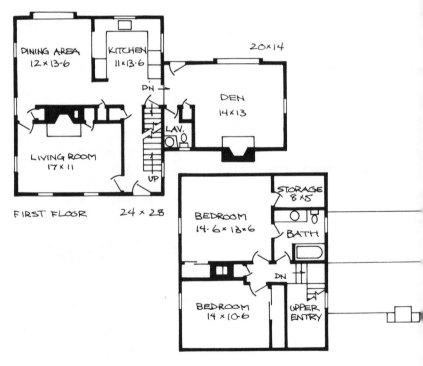

Monterey Horizon 20

Type of Structure: Shell package for 20' geodesic dome

Manufacturer: Monterey Domes
P.O. Box 55116
Riverside, CA 92517
(714) 684-2601

Materials Provided: All framing members are kiln-dried Douglas fir. Package includes struts and studs for 60 triangles plus the steel hubs to join them. Covering is ½" exterior-grade plywood. Also included in this kit are standard 4' riser wall components. All pieces are precut, predrilled, and color-coded. Hardware consists of bolts, nuts, washers, and nails of heavy-duty steel. Model has five ground-floor openings; extension shown here not standard. Complete engineering drawings also included.

Exterior Dimensions: 20' diameter, 12' high

Living Area: 350 sq. ft.

Price: $3,995 FOB Riverside, CA

Price/sq. ft.: $11.41

Warranty: 90 days for materials and workmanship

Special Features: This kit is not as fully prefabricated as some, but the company feels its color-coding system makes construction possible for the inexperienced. The heaviest piece weighs less than 30 lbs. A wide range of options, including 2×6 framing rather than the standard 2×4, skylights, wooden shakes and shingles, synthetic vinyl shakes for weathertight roofing (heavily insulated with polyurethane and fire-resistant), higher base walls, extensions, and rigid polyurethane insulation. Domes from 20' to 45' in two styles; also clusters.

Energy Features: Monterey Domes estimates fuel savings of 30-50%

for its dome. Firm offers solar collectors and systems through tie-in with independent manufacturer.

Notes: "You need only a hammer, wrench, ladder and level to make your dome frame a speedy reality," says Monterey Domes. They estimate that one unskilled worker can erect the frame in a few days. Shipping cost for Horizon 20 from Riverside to New York less than $300.

Price of Information Kit: $6

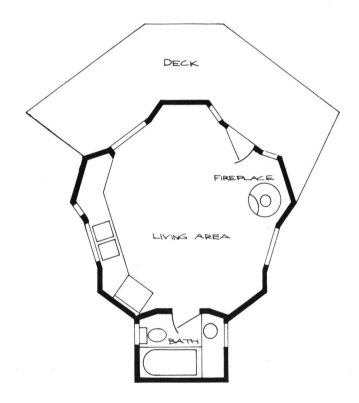

Heritage Log Greenbrier

Type of Structure: Two-story partial log shell package

Manufacturer: Heritage Log Homes, Inc.
P.O. Box 610
Gatlinburg, TN 37738
(615) 436-9331

Materials Provided: Exterior log walls of precut 8"-thick Southern yellow pine, pretreated, smooth-milled, and numbered. Maximum length 10'. Logs are double tongue-and-groove with foam insulation strips and saddle gaskets for interlocking corners. Insulated splines and 10" steel spikes for assembly. Prehung windows with double-insulated glass and wood sash; prehung exterior doors of solid pine or fir with oak threshold. Milled log porch posts included. Owner furnishes roof rafters and covering, floor joists and flooring, stairs, interior partitions, gable-end framing, and all interior finish.

Exterior Dimensions: 35×25'

Living Area: Upper   289 sq. ft.
Lower   875 sq. ft.
Total   1,164 sq. ft.

Price: $9,700 FOB Gatlinburg, TN

Price/sq. ft.: $8.33 for partial shell only

Warranty: 90 days for materials

Special Features: Options include Thermopane windows, log gable ends, rough-sawn timber rafters and second-floor framing package, dormer windows, garage. Heritage Log Homes offers 30 standard models and ships throughout the East, Midwest, and South. "No messy caulking necessary," states company; uniformly milled solid pine logs mean no ledges or uneven joints to catch and hold water.

Notes: Shipped on 40' flatbeds; three or four people recommended for unloading; forklift can be used. Packed to eliminate log-sorting tasks. Greenbrier requires one truck, and average construction time for package for an experienced crew of three or four is less than a week, Heritage says.

Price of Information Kit: $5

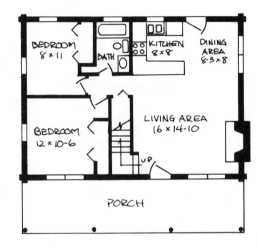

FIRST FLOOR   35×25

SECOND FLOOR

Nor-Wes Cedar Cowichan

Type of Structure: Two-story complete shell package

Manufacturer: Nor-Wes Cedar Homes
11120 Bridge Road
Surrey, British Columbia
Canada V3V 3T9
(604) 580-2024

Materials Provided: Floor system of Douglas fir joists and plywood subflooring; upper floor same with main-floor ceiling of tongue-and-groove Western red cedar decking. Exterior walls of fir studs, fiberglass insulation, building paper and rough-faced channel cedar siding. Interior wall finish for all partitions and exterior walls Western red cedar. Roof system of fir rafters, tongue-and-groove red cedar decking, rigid insulation, felt, strapping, split-cedar shakes. All major components precut. Exterior doors prehung and weatherstripped solid cedar; interior doors stain-grade rotary mahogany, knocked down. Windows and patio doors wood sash, double glass with screens. Precut interior stairs with railings; lower exterior decks and upper balconies include flooring, posts, and railing of fir. Miscellaneous hardware. Same package available with spruce, pine, fir finish for interior walls, roof decking and ceiling of main floor; $47,795 for spruce interior.

Exterior Dimensions: 50×18' plus decks

Living Area: Upper 640 sq. ft.
Lower 1,020 sq. ft.
Total 1,660 sq. ft.

Price: $53,120 FOB Surrey, B.C., in Canadian dollars

Price/sq. ft.: $32

Warranty: None

Special Features: A complete insulated-shell package. Duty and brokerage fees for American customers are about 4½-5% of total cost but replace local and state taxes; in addition, as of fall 1981, American dollars were worth 22% more than Canadian. Options include interior wall finish or resawn clear cedar, special entry doors, and custom trim.

Energy Features: Walls and roof are R-20.62, with more insulation available. Design of A-line chalet featured here is ideal for solar collection system. Triple-glazed windows optional.

Notes: "A good percentage of the people who buy our homes do build them themselves," says Nor-Wes. Shipped by rail.

Price of Information Kit: $6

SECOND FLOOR

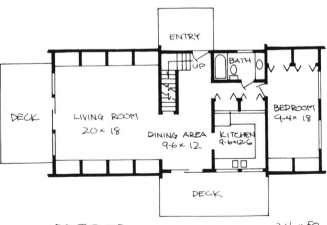

FIRST FLOOR 24 × 50

Lok Log Powderhorn

Type of Structure: One-story log shell package

Manufacturer: Lok Log
Building Logs, Inc.
P.O. Box 300
Gunnison, CO 81230
(303) 641-1844

Materials Provided: Logs are 7¼" lodgepole pine, milled from dead or dry trees and factory-treated with preservative, precut, drilled, and numbered. Logs are beaded for acrylic latex caulking and foam weatherseal for all seams, butt joints, and corners. Wooden 4' dowels and 8½' steel bolts tighten and locate walls. Fir framing for floor. Precut T jambs for windows and fireplaces, top and bottom plates, precut floor system of girders, joists, sill plates, bridging and blocking, plywood. Roof system of prebuilt trusses or log rafters (depending on model), plywood, felt, asphalt shingles, frieze blocks, fascia, and shingle bead. Gables prebuilt of reverse board-and-batten plywood. Pine framing for interior walls. Prehung and cased exterior doors of fir; cased windows of insulated glass with screens. Miscellaneous hardware and log oil. Owner furnishes insulation and interior finish materials.

Exterior Dimensions: 38½ × 24½'

Living Area: 943 sq. ft.

Price: $17,464 FOB Gunnison, CO. Free delivery within 350 miles of Gunnison.

Price/sq. ft.: $18.52

Warranty: "Delivery to customer of sound, good quality, properly milled logs"

Special Features: For its 33 models (most with variations), Building Logs offers your choice of a shell kit only, a shell with interior materials, and a kitchen-cabinet kit, plus a variety of balconies, porches, dormers, etc. Plans can be customized.

Notes: Kits of $30,000 or less can be hauled in one load to accessible site. Charge from Gunnison to Chicago about $1,995; to Atlanta about $2,565. Says Lok Log literature, "Two to 3 men can generally construct the shell of a two bedroom home in one or two weeks."

Price of Information Kit: Free

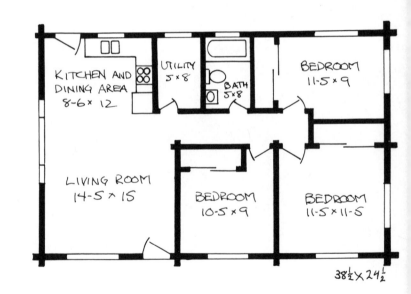

Barn People's Hay Barn, Granary, and Horse Barn

Type of Structure: Restored and reassembled original barn frame

Manufacturer: The Barn People
P.O. Box 4
South Woodstock, VT 05071
(802) 457-3943

Materials Provided: Blueprints of suggested foundation and first-floor deck for owner's site preparation; color photo of barn on original site and during dismantling; balsa scale model of frame; blueprints of frame; history and architectural survey of barn. The Barn People deliver and erect (with crane) the basic frame of the barn, which is fastened with wooden pegs. The entire frame has been carefully labeled, dismantled, and checked for soundness, with timbers repaired or replaced, test-assembled, and wirebrushed prior to delivery. Offered at extra charge is an adaptation service for alterations to original frame configuration. Each structure is one of a kind.

Exterior Dimensions: Hay barn 20×30', granary 16×18', horse barn 20×20'

Living Area: Upper   981 sq. ft.
Lower 1,384 sq. ft.
Total   2,365 sq. ft.

Price: Prices range from $10,000 to $40,000

Warranty: None

Special Features: No two barns are alike, so *The Barn People have no standard floor plans.* Many interior and exterior finishes are possible, as are floor level and partition and door/window positioning. Inventory folders of available barns are compiled quarterly; prospective clients are urged to visit standing barns. Options are offered only for the sake of authenticity and include sheathing with honey-colored random-width boards for roof, lofts, and levels; ladders; framing studs for windows and doors; additions such as sheds and ells; barn doors; yellow pine floorboards, barn board, and beams. All materials are carefully matched to the barn's interior. Shipping for options is extra.

Notes: The Barn People recommend that you work with your own architect and contractor for major tasks. Delivery schedules may be booked for 6-10 months in advance, so early ordering is advised. Frames are unloaded from flatbed trucks. Shipping charges from Windsor, VT, range from $4 to $6 per loaded mile.

Price of Information Kit: $10

Northern Glen Spey

Type of Structure: Two-story panelized shell package

Manufacturer: Northern Homes, Inc.
10 LaCrosse Street
Hudson Falls, NY 12839
(518) 747-4128

Materials Provided: Sills and sealer, girders, floor joists of spruce or hemlock, precut bridging, plywood subfloor. All structural lumber is Western kiln-dried. Exterior wall is wafer board over fir studs with siding of Masonite, aluminum, clapboard, board-and-batten, V–groove, or several other options. Roofing system of white fir joists and rafters, plywood sheathing, felt and asphalt shingles standard. Exterior doors prehung and insulated core steel; interior doors prehung flush birch; windows (with screens) of welded insulated glass. Patio doors tempered insulated glass with screening. Underlayment; flooring of select red oak. Insulation unfaced R-13 for walls with poly vapor barrier, faced for ceilings, R-38. Soffits, fascia, louvers, basement stairs precut, moldings, miscellaneous trim and hardware also included.

Exterior Dimensions: 32×42' with decks

Living Area: Upper   318 sq. ft.
Lower   1,008 sq. ft.
Total   1,326 sq. ft.

Price: $25,764 for shell package; estimated turnkey cost $65,000. Northern Homes feels do-it-yourselfers can save 20-30% depending on skill and ambition. Price includes delivery.

Price/sq. ft.: $19.43 for shell; $49.02 for turnkey

Warranty: Northern guarantees a number of its items for varying periods of time, some for the life of the home.

Special Features: Another quality outfit, Northern has been doing business since 1946 and has a network of builder/dealers. A wide range of designs and materials is offered, and components are first-rate. Packages are more complete than most and are available throughout New England and the mid-Atlantic states.

Energy Features: A booklet of special tips and good energy-conservation advice is included. Also, good attention to detail for weathertight doors, windows, and ceilings.

Notes: Do-it-yourselfers can assist and do finish work, good construction guide. Shipped by flatbed; no special unloading equipment.

Price of Information Kit: $5

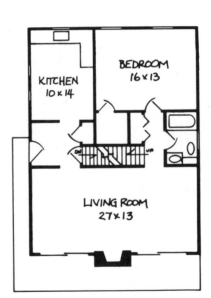

FIRST FLOOR   32 × 42

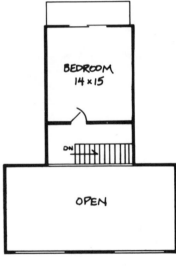

SECOND FLOOR

National Beauti-Log 1000

Type of Structure: One-story log shell package

Manufacturer: National Beauti-Log Cedar Homes
P.O. Box 6374
Stockton, CA 95206
(209) 465-3437

Materials Provided: Exterior walls are 4×8 double tongue-and-groove unseasoned Western red cedar. Grooves are tapered and corners have interlocking joints; caulking supplied for corners. Logs are milled for finished exterior and interior sides. Floor beams of Douglas fir, white fir subflooring, particle-board underlayment. Douglas fir purlins, roof decking of cedar (optional) or knotty pine tongue-and-groove, felt, asphalt shingles. Interior walls and partitions same as exterior. Prehung aluminum single-glazed windows with screens standard. Exterior doors solid-core, interior hollow-core, prehung. Miscellaneous interior trim, flashing, and hardware. Owner supplies roof insulation.

Exterior Dimensions: 39×26′

Living Area: 1,014 sq. ft.

Price: $24,765 FOB Stockton, CA

Price/sq. ft.: $24.42

Warranty: Will replace miscuts and shortages within 10 days for items other than logs; for logs, 30 days after delivery.

Special Features: Interior walls and roofing make this a more complete package than some. Wall logs lock together, says manufacturer, for a waterproof and airtight seal without nails, pins, or dowels. Red cedar resists rot and insects. Optional items are double-glazed windows (wood or aluminum) and cedar shakes, roof insulation.

Notes: Shipped on 40′ trucks; forklift or three or four people recommended for unloading. Sledgehammers needed for construction. With help, walls can be up in a matter of days.

Price of Information Kit: $1

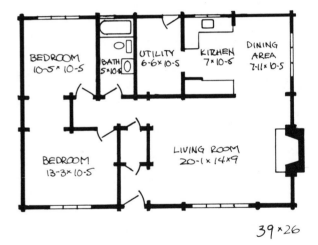

39×26

Geodesic Sierra

Type of Structure: Panelized shell package for 39' geodesic dome

Manufacturer: Geodesic Domes, Inc.
10290 Davison Road
Davison, MI 48423
(313) 653-2383

Materials Provided: Sixty predrilled triangular panels of kiln-dried Southern yellow pine 2×4s (2×6s optional) covered with 5/8" resin-board sheathing, secured by metal clips. All hardware and detailed prints included. This model has five openings.

Exterior Dimensions: 39' diameter, 16'8" high

Living Area: Upper   800 sq. ft.
Lower 1,100 sq. ft.
Total   1,900 sq. ft.

Price: $6,800 FOB Davison, MI

Price/sq. ft.: $3.58

Warranty: One year

Special Features: Geodesic Domes has been making domes for more than 24 years. Options include riser walls, dormers, extensions, and skylights. Kit domes range from 26' to 45' and can be clustered. Company does commercial buildings and churches up to 110' in diameter.

Energy Features: Savings of up to 50% on fuel bills claimed for these domes. Recommend urethane foam insulation with overlay of shingles on outside of dome for permanent residences.

Notes: Estimated construction time for this shell is two days for unskilled builders. Hammers, wrenches, and light scaffolding can be used. Commercial shipping or can be picked up at the factory in a rented truck.

Price of Information Kit: $8

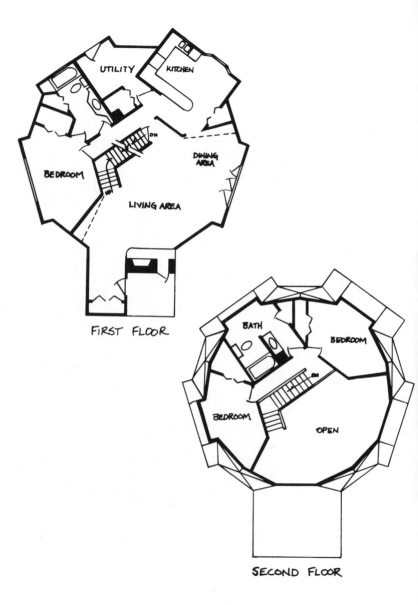

FIRST FLOOR

SECOND FLOOR

Solid Wall Sentry Saltbox

Type of Structure: Two-and-a-half-story complete beam-shell package

Manufacturer: Solid Wall Buildings, Inc.
P.O. Box 41
Rts. 11 & 103
Newport, NH 03773
(603) 863-3107

Materials Provided: Conventional first-floor deck system with plywood subflooring. Exterior walls of 4×6" Eastern white pine, precut and numbered; clapboard exterior, 2" of urethane insulation; V-groove pine panel effect on the interior. Alternating interlocking tight joints at corners. Main beam 8×12", ceiling joists 4×6", 2' on center (all precut, notched, and lap jointed). Conventional 2×10 rafter system with plywood sheathing and asphalt shingles. Complete exterior trim package. Optional tongue-and-groove decking for second and third floors. Prehung cased door units and double-hung, insulated tilt-out windows with screens.

Exterior Dimensions: 36×36'

Living Area:
Third floor	540 sq. ft.
Second floor	1,080 sq. ft.
First floor	1,296 sq. ft.
Total	2,916 sq. ft.

Price: $34,000 FOB Newport NH

Price/sq. ft.: $11.66

Warranty: For materials and workmanship, one year

Special Features: Solid Wall offers custom designing for styles ranging from New England traditional to contemporary. Their houses combine post and beam with log construction techniques and use a variety of exterior finishes.

Energy Features: In February 1980 the local utility company calculated heat loss for the Sentry Saltbox and estimated that it would cost $1,720 to heat over a 12-month period. The actural cost of heating this large house was only $795 (temperature set at 68°). Solid Wall attributes this better-than-50% energy savings to the thermal storage capacity of the wood walls (which also have an R-value of 25) and the interior shutter system (which increases the R-value of the windows from 2 to 12). For additional savings, passive and active solar systems and other conservation measures can be incorporated in the house design.

Notes: Solid Wall says that its "kit package is truly designed for the do-it-yourselfer or someone trying to cut labor costs." Shipped nationwide on 40' trailer.

Price of Information Kit: $5

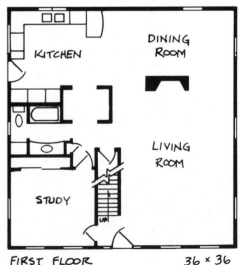

FIRST FLOOR   36 × 36

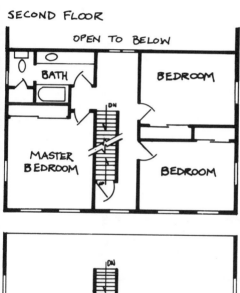

SECOND FLOOR

OPEN TO BELOW

BATH

BEDROOM

DN

MASTER
BEDROOM

BEDROOM

DN

THIRD FLOOR

Green River Saltbox

Type of Structure: Two-story log shell package

Manufacturer: Green River Trading Co.
Boston Corners Road
Millerton, NY 12546
(518) 789-3311

Materials Provided: Species of log depends on mill location: Massachusetts mill uses Eastern white pine, Colorado and Oregon mills furnish Engelmann spruce or lodgepole pine. All wall logs (including gable ends) are double tongue-and-grooved with round exterior and flat interior surfaces. Standard log dimensions 6×6½″ with options of wider widths; round interior wall also available. Sealing between logs done with closed-cell polyethylene gasket material, 10″ spikes, and acrylic-rubber caulking for all joints. Rough-sawn dimensional timbers of hemlock, spruce, or fir for structural members: porch posts and top plates; second-floor joists, girders, and girder posts; tie beams; porch rafters; balcony posts; collar ties; purlins; and braces. Clear wood preservative included. Owner provides roof and rafter system, doors, and windows. Roof rafters and ridge beams available from GRTC at nominal additional cost.

Exterior Dimensions: 34×24′ with 34×6′ porch

Living Area: Upper 816 sq. ft.
Lower 428 sq. ft.
Total 1,244 sq. ft.

Price: $10,260 FOB Great Barrington, MA; Eugene, OR; or Salida, CO

Price/sq. ft.: $8.25

Warranty: One year against defects in materials and workmanship

Special Features: Wall logs flat on interior side for flexibility of finish; round interior side extra cost, as are optional 8″, 9″, or 10″ logs. Green river has four basic models ranging from $6,950 to

$10,260 as well as garages, barns, and other structures.

Notes: By eliminating packaged doors, windows, and roof system, Green River is able to get its kit on one tractor trailer to minimize costs. Shipping costs minimum $150 for first 100 miles plus $1.25 per mile thereafter. Company suggests that builders without much carpentry experience seek professional help for the major tasks.

Price of Information Kit: $5

FIRST FLOOR   34 × 24

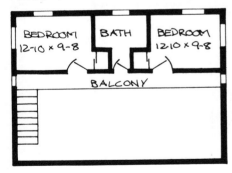

SECOND FLOOR

Pre-Cut International Belaire

Type of Structure: Two-story solid laminated wall system

Manufacturer: Pre-Cut International Homes
P.O. Box 97
Woodinville, WA 98072
(206) 668-8511

Materials Provided: Tongue-and-groove laminated Western red cedar for exterior walls, fir floor beams, plywood subfloor, tongue-and-groove decking with particle-board overlay for loft floors. Roof tongue-and-groove decking, polystyrene with furring strips, felt, shake liner and hand-split cedar shakes. All trim for exterior and interior supplied. Double-glazed insulated windows and sliding doors with screens, doors of flush mahogany, all prehung. Hardware, caulking, adhesives, flashings included.

Exterior Dimensions: 30×36'

Living Area: Upper 258 sq. ft.
Lower 1,080 sq. ft.
Total 1,338 sq. ft.

Price: Standard wall $37,308; Nominal 5, $43,490; Thermo-Lam in cedar, $44,418 (less in pine). FOB Woodinville, WA.

Price/sq. ft.: Standard, $27.88; Nominal 5, $32.50; Thermo-Lam, $33.20

Warranty: One year for materials

Special Features: Exterior walls joined with patented T joint for extra strength; factory-applied preservative for exterior wood. Each model (there are about 40) comes in three versions: Standard wall (two laminations of cedar with Douglas fir or hemlock in center); Nominal 5 (five laminations of cedar); Thermo-Lam (two laminations of cedar with polystyrene insulation in center). Houses meet or exceed VHA and FHA requirements, as well as most local and state zoning requirements.

Notes: Pre-Cut International literature states that more than half their buyers participate in the construction of their homes. Shipped on one or more flatbeds; forklift needed for unloading.

Price of Information Kit: $5

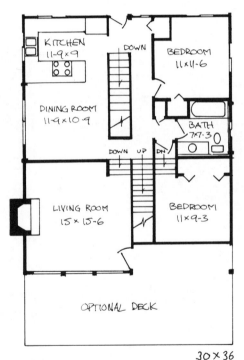

SECOND FLOOR

FIRST FLOOR

30 × 36

150

Four Seasons Niagara

Type of Structure: One-story log shell and interior package

Manufacturer: Four Seasons Log Homes
3425 Major McKenzie Drive
Woodbridge, Ontario
Canada L4L 1A6
(416) 832-2945

Materials Provided: Kit available in three packages. Phase 1 includes floor system of sill plates, beams for interior piers, joist system, 5/8'' tongue-and-groove plywood subfloor, bridging, caulking, base flashing, pine base trim. Wall system of precut, notched, and coded 3'' pine logs. Roof system of fir beams, 2×6'' pine roof boards, heavy-duty asphalt shingles, flashing, pine soffit and fascia. Interior partitions; double-glazed windows with screens; cedar exterior doors; mahogany interior doors (others available); closets with doors; insulation/packing for use around openings; nails; caulking. Phase 2 package includes above specifications and adds vacuum-sealed insulated picture windows, wood storm/screen doors, insulated roof (R-20 fiberglass). Phase 3 includes Phase 2 specifications except to substitute double wall and roof systems. Double wall of 3×6'' pine logs with 3½'' R-12 fiberglass and vapor barrier in cavity between them. Double roof of R-32 fiberglass insulation, vapor barrier, 1×6'' tongue-and-groove pine inner decking, 7/16'' aspenite sheeting, self-sealing asphalt shingles, valley flashing.

Exterior Dimensions: 50×27'

Living Area: 1,319 sq. ft.

Price: Phase 1, $32,090; Phase 2, $35,670; Phase 3, $43,790 (Canadian dollars) FOB Parry Sound, Ontario

Price/sq. ft.: Phase 1, $24.33; Phase 2, $27.04; Phase 3, $33.20 (Canadian dollars)

Warranty: Any defective materials shipped will be replaced.

Special Features: Main entrance foyer with closet; large living room; spacious kitchen with breakfast bar. Plans can be customized.

Energy Features: Four Seasons stresses the energy efficiency of their homes. The solid log walls provide continuous insulation and there is no breaking point in the roof. The Phase 3 package provides an outstanding superinsulated home.

Notes: Kit is delivered on flatbed truck and pup trailer or by rail. Everything rolls off the truck; no special equipment is needed. Four Seasons says that the house is 25% finished when delivered and that 35% of their customers build their own homes. Construction time, three to four weeks.

Price of Information Kit: $2

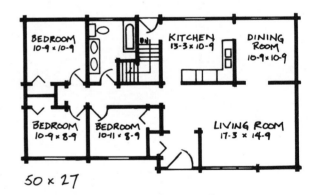

50 × 27

Kingsberry Tahoe II

Type of Structure: One-and-a-half story panelized complete shell package

Manufacturer: Kingsberry Homes
The Boise Company
P.O. Box 8358
Boise, ID 83707
(208) 344-3113
(See regional addresses below.)

Materials Provided: Choice of foundation. Precut and numbered floor system of sills, sealers, joists, plywood subfloor with underlayment, fiberglass insulation, moisture barrier. Walls are preassembled sections with single-glazed windows and redwood beveled siding standard. All framing precut, insulated sheathing and fiberglass insulation for walls and roof. Preassembled gables, intrerior wall assemblies, and partitions. Roof system or trusses or precut framing, plywood decking, felt, and asphalt shingles with thermal baffles and soffit vents. *This package must be erected by a local builder,* but with Kingsberry's Home Earner Program you can subcontract part of the interior work and also do some yourself, depending on your skills.

Exterior Dimensions: 26×40'

Living Area: Upper   303 sq. ft.
Lower   985 sq. ft.
Total   1,288 sq. ft.

Price: Ranges from $51,000 to $62,000 depending on options selected, geographical location, and amount of work by buyer. Includes foundation.

Price/sq. ft.: $40-$48

Warranty: Offered to builder

Special Features: Kingsberry Homes has been around since 1946. This is a complete package sold to independent builders from plants in Oklahoma, Virginia, and Alabama. A wide range of options for almost 200 designs and good materials along with builder and manufacturer support mean less hassle at the bank. The Boise Company markets a sectional home in the Western states.

Energy Features: Insulation is R-13 for floor and walls, R-19 for ceilings standard. Double and triple glazing of windows optional, as are storm doors and windows and a number of additional energy savers.

Notes: "A typical Kingsberry home can be erected in a matter of days" by the builder, says The Boise Company. Once the shell is dried in, the do-it-yourselfer can arrange for utilities and finish work as time and budget allow.

Price of Information Kit $5

In DC, VA, WV, NC, SC, OH, MI, PA, MD, DE, NJ, NY, ME, NH VT, MA, CT, and RI contact:
Kingsberry Homes
P.O. Drawer "B"
U.S. Highway #58
Emporia, VA 23847
(804) 634-2154

In AL, GA, FL, MS, AR, LA, TN, and KY contact:
Kingsberry Homes
P.O. Box 228
1725 Gault Avenue South
Fort Payne, AL 35967
(205) 845-3550

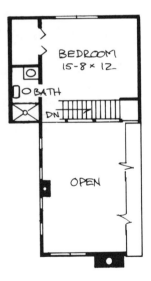

BEDROOM
15-8 × 12

BATH

DN

OPEN

SECOND FLOOR   21-1 × 16-4

In NM, TX, OK, CO, KS, and
also AR and LA, contact:

Kingsberry Homes
P.O. Box 938
201 North Kingsberry Road
Holdenville, OK 74848
(405) 379-5437

BEDROOM
12-9 × 9-8

BEDROOM
9-10 × 12

BATH

KITCHEN

UP

LIVING ROOM
15-8 × 19-8

DINING AREA
9-5 × 12-4

OPTIONAL DECK

FIRST FLOOR   26×40

Vermont Log Hartland

Type of Structure: Two-story log shell package

Manufacturer: Real Log Homes
Vermont Log Buildings, Inc.
Hartland, VT 05048
(802) 436-2121

Materials Provided: Type of log depends on location of plant; Vermont plant uses Eastern white pine, North Carolina and Arkansas plants furnish Southern yellow pine, and Montana and Nevada plants provide lodgepole pine. Diameters are 8-10″ with maximum lengths up to 12′. Sealing with Lockspline, 10″spikes, and polyvinyl chloride foam gaskets. Doors and windows are prehung, insulated windows and storm-screen combinations options. Log girders and ceiling joists, rafters, ridgepoles provided as well as porch rafters, posts, sill and plates, balcony-railing stock, and fireplace mantel. Owner provides first-floor joists and girders, interior partitions, stairs, underlayment, finished flooring and roof covering, foundation, and fireplace.

Exterior Dimensions: 32×16′

Living Area: Upper 261 sq. ft.
Lower 626 sq. ft.
Total 887 sq. ft.

Price: $13,100 FOB plant

Price/sq. ft.: $14.77

Warranty: Repair or replace within 120 days

Special Features: Model in photo has a Cavendish dining extension at extra cost. Optional second-floor girder and joist system permits an additional 268 sq. ft. for another bedroom.

Notes: Shipped on 40′ flatbeds; at least three persons required for unloading. "An experienced crew of four can erect the walls, doors, windows, floor joists and roof system of most models in a matter of weeks rather than months," says Vermont Log.

Price of Information Kit: $5

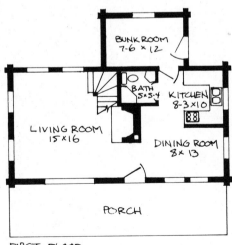

FIRST FLOOR

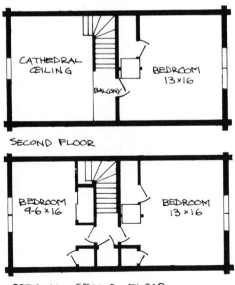

SECOND FLOOR

OPTIONAL SECOND FLOOR

Rondesics R-15

Type of Structure: One-story panelized shell package

Manufacturer: Rondesics Homes Corp.
527 McDowell Street
Asheville, NC 28803
(800) 438-5859
(704) 254-9581 in NC

Materials Provided: Floor beams; floor panels of 2×6 framing, R-19 fiberglass insulation, and plywood; particle-board underlayment. Wall system of 15 8'×8' panels, each with stud framing, superthick R-19 fiberglass insulation, redwood or cedar siding. Double-glazed wood casement windows and steel-clad exterior doors already installed in wall panels. Roof system of precut rafters, plywood sheathing, 9" R-30 fiberglass insulation, felt, and asphalt shingles. Roof overhang and fascia of heart redwood or cedar; rough-sawn redwood or cedar soffits. Decking, rails, trim, and posts of heart redwood or cedar. All hardware, nails, and spikes included. Options include interior packages, extra deck sections, ramps, exterior stairs, steel columns, special wall sections (for fireplace, double doors, etc.), loft. Model shown has extra panels to create basement area.

Exterior Dimensions: 39'3" in diameter

Living Area: 1,202 sq. ft.

Price: $26,452 for package outlined above; lower-priced packages available

Price/sq. ft.: $22.01

Warranty: One year for materials and workmanship; five years for windows

Special Features: Rondesics' structures are adaptable to many exterior and interior arrangements. Any of the six basic models (ranging from 300 sq. ft. to more than 2,100 sq. ft) can be combined, and like sizes can be stacked. Interior walls are neither preconstructed nor load-bearing, so they can be placed anywhere the buyer prefers. The floor plan shows only one of many possible interior designs that can be worked out by Rondesics Homes' custom design service.

Energy Features: Because less wall space is exposed to the elements, a Rondesics home can save on heating and cooling costs over a comparably sized rectangular home constructed of similar materials. When properly oriented on a site these homes are naturally good solar collectors. Many owners also use solar panels.

Notes: Shipped by closed trailer; company will compute freight costs. No special equipment needed for unloading.

Price of Information Kit: $3

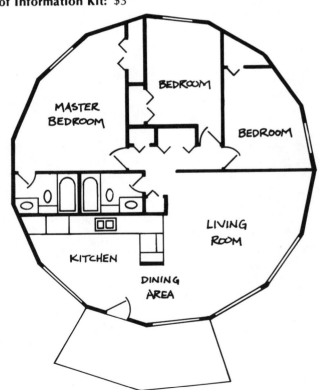

Timber Kit No. 2

Type of Structure: Two-story post-and-beam shell package

Manufacturer: Timber-Kit by Habitat
123 Elm Street
South Deerfield, MA 01373
(413) 665-4006

Materials Provided: Precut, prenotched Douglas fir frame, Western pine tongue-and-groove decking for upper floor, roof and optional loft. Walls preassembled, kiln-dried with factory-framed window and door openings. Insulation is 1" styrofoam for walls, 3" urethane for roof — owner supplies fiberglass. Double-hung insulated windows; prehung exterior doors with foam-insulated core. Roof shingles extra-weight, textured. Kiln-dried siding with preservative stain for exterior walls. All hardware and trim provided; air-lock entry.

Exterior Dimension: 24×28'

Living Area: Upper   640 sq. ft.
Lower   640 sq. ft.
Total   1,280 sq. ft.

Price: $23,263 FOB Amherst, MA, shipping about $1 per mile. Deductions can be made if you wish to provide your own shingles, decking, insulation, siding, windows, and doors.

Price/sq. ft.: $18.17 for complete shell package

Warranty: One year for materials

Special Features: Optional loft and skylights. Timber-Kit offers 18 models ranging in price from $20,929 to $30,886.

Energy Features: Wide range of energy-conserving ideas in all Timber-Kit packages. House is ready for solar or conventional heating system, or solar can be added later both for space heating and hot water. With recommended insulation on walls, roof, and footing and under slab, R-values of 22, 26, and 8.6 are claimed respectively. Homes designed for siting with minimum north-facing openings and maximum glass on southern exposures. Optional solar greenhouses.

Notes: Timber-Kit packages are the result of this company's experience with post-and-beam structures through their Habitat line. Owner participation is encouraged.

Price of Information Kit: $2

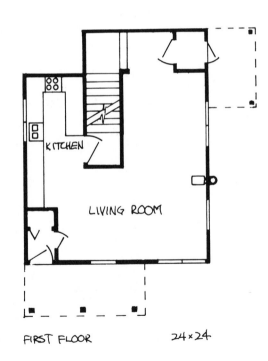

FIRST FLOOR   24×24

160

BEDROOM

BATH

BEDROOM   BEDROOM

SECOND FLOOR

Alta Walton

Type of Structure: One-story log shell package

Manufacturer: Alta Industries, Ltd.
P.O. Box 88
Halcottsville, NY 12438
(914) 586-3336

Materials Provided: Exterior walls of solid white pine logs, almost 5″ thick, precut and coded, maximum length 16′. Log system of double grooves top and bottom, splines, and caulking; 12″ spikes. Logs are planed inside and out and joints interlock. Precut and numbered gables, snow blocks, laminated yellow pine or fir ridge beam, Douglas fir rafters and plywood decking. Also includes insulated metal exterior door, wood casement windows, miscellaneous hardware, and caulking. Owner supplies interior partitions, floor system, roofing, insulation and sill seal, utilities, and other finish.

Exterior Dimensions: 24×35′

Living Area: 720 sq. ft.

Price: $9,923 FOB Halcottsville, NY

Price/sq. ft.: $13.78

Warranty: 10 years on material and workmanship

Special Features: Alta will also construct the house for you. This company offers more than 30 models and a variety of sizes in a price range from $8,453 to $30,345. Advice on obtaining a mortgage is provided. About 55 dealers in 18 states. Alta also builds commercial structures, including the ski lodge used for the 1980 Winter Olympics in Lake Placid.

Notes: Because this package is not a complete house, it can be shipped on one truck rather than two. Three or four people are needed for unloading; experienced crew can dry in shell in about two days.

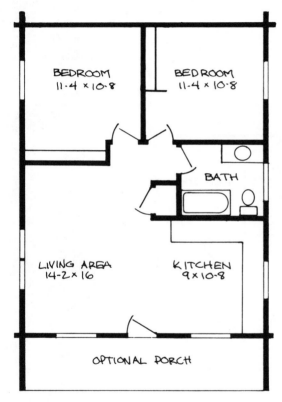

24 × 35

Birch Hill Saltbox

Type of Structure: Pine post-and-beam frame for one-and-a-half-story saltbox with ell

Manufacturer: Birch Hill Builders
Brave Boat Harbor Road
York, ME 03909
(207) 363-2814

Materials Provided: Birch Hill provides frames of white pine with oak pegs. Frames can be precut, or the firm will provide plans from which clients can do their own cutting. Rough-sawn timbers cost $4.70 per square foot; major timbers can be hand-planed for $5.50, or all timbers can be hand-planed for $6.50. If Birch Hill raises the frame, the cost is an additional $1.30 per square foot, approximately. Frame pictured here is for a slightly different saltbox.

Exterior Dimensions: 28×32' with 10×10' ell

Living Area: Upper   576 sq. ft.
Lower   996 sq. ft.
Total   1,572 sq. ft.

Price: $9,468 for rough-sawn package, FOB York, ME

Price/sq. ft.: $4.70 rough-sawn, frame only

Warranty: If Birch Hill raises frame, "we guarantee it for 200 years."

Special Features: Birch Hill builds frames for Capes, saltboxes, barns and garrison-style houses, and "will provide anything from a timber order list, frame drawings, measured timber joinery drawings, color-coded raising instructions, conversion of conventional plans to timber frame plans, incorporation of solar—both passive and active—into the design."

Notes: Shipping from York, ME, at $1.50 per mile one way (less for longer distances). *The Timber Framing Book* (Housesmiths Press, Kit-tery Point, ME $11.95 plus .75 postage) is recommended for those interested in traditional 18th-century framing techniques.

Price of Information Kit: $4

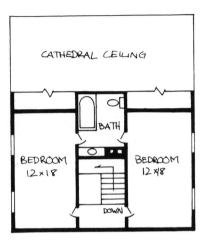

SECOND FLOOR

FIRST FLOOR   28 x 32

Pan Abode Islander I

Type of Structure: One-story log shell package

Manufacturer: Pan Abode, Inc.
4350 Lake Washington Blvd., N.
Renton, WA 98055
(206) 255-8260

Materials Provided: Wall system of air-dried Western red cedar timbers, 3×6″ or 4×6″ nominal dimensions, single tongue and groove with interlocking cross joints (exterior and interior walls). Floor system of 2×10 hemlock/fir floor joists, plywood subfloor, particle-board underlayment. Roof system of precut gable timbers, #1 Douglas fir beams, 2×6 hemlock tongue and groove decking, 3½″ rigid insulation, plywood sheathing, felt, asphalt shingles (cedar shakes optional). Prehung raised-panel Western red cedar interior and exterior doors. Wood-framed insulated sliding windows with screens. Miscellaneous hardware, interior and exterior trim, caulking, and stain. Complete drawings and 68-page construction manual.

Exterior Dimensions: 34×22′

Living Area: 738 sq. ft.

Price: $19,250 (4×6 wall timbers) FOB Renton, WA

Price/sq. ft.: $26.08

Warranty: One year

Special Features: Cedar shakes, skylights, cedar decks, "Energy Wall" insulation package, triple-glazed windows, garage available as options. Thirty-four other one- and two-story house models are offered, as are duplexes, townhouses, and condominiums. Complete custom designing is also available.

Notes: Kits are shipped by flatbed truck, van, or containers (to Alaska, Hawaii, and overseas). A forklift or boom crane is required for unloading. Company says these homes are "designed for owner assembly." Construction time estimated at an average rate of 100 sq. ft. per day using two workers and ordinary carpentry tools.

Price of Information Kit: $10

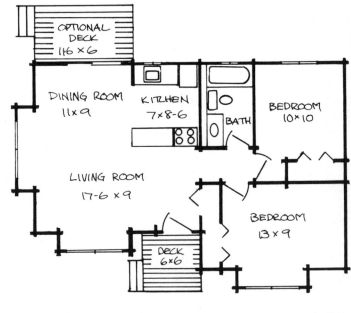

Cluster Shed #1

Type of Structure: One-story post-and-beam shell package

Manufacturer: Timberpeg
Box 1500
Claremont, NH 03743
(603) 542-7762

Materials Provided: Eastern white pine timbers with interlocking mortise-and-tenon joints, pegged with square oak trunnels for frame. Exterior walls resawn pine board-and-batten; interior white vinyl-faced Homasote. Roof and wall systems of tongue-and-groove kiln-dried pine, 2″ rigid isocyanurate insulation with both sides foil-faced, strapping. Shingle ribs, hand-split Western red cedar shakes for roofing. Double-pane insulated windows, pine doors, patio doors of tempered insulating glass. Hardware and trim included.

Exterior Dimensions: 16×12′

Living Area: 192 sq. ft.

Price: $5,372 FOB Claremont, NH; Elkin, NC; or Aurora, CO. Includes shell only, no interior finish.

Price/sq. ft.: $27.98

Warranty: 30 days for quantity and workmanship

Special Features: This is the simplest of four basic Timberpeg house packages designed to be combined with each other or larger models. Cluster Shed prices range from $5,372 for single units and up to $60,000 for combinations. Highly adaptable to sloping sites.

Energy Features: System includes layer of tongue-and-groove and layer of 2″ insulation board separated by strapping to create air space. Insulation values said to be R-21 for walls and R-20 for roof. Optional 3″ insulation.

Notes: Timberpeg materials are good quality, details are attractive, and simpler designs are possible for do-it-yourselfers. Network of independent sales representatives or consultation with manufacturer suggested. Shipped on 40′ flatbed trucks.

Price of Information Kit: $10

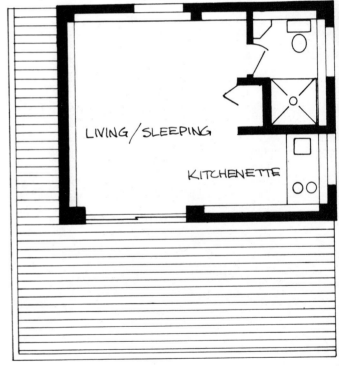

16 × 12

Bellaire Lakeside

Type of Structure: One-and-a-half-story log home package

Manufacturer: Bellaire Log Homes
Box 322K
Bellaire, MI 49615
(616) 533-8633

Materials Provided: Exterior walls of air-dried Northern white cedar 8' half-logs pretreated and smooth-finished. Precut logs joined vertically with wood splines and 8" spikes. Whole-log rafters and ties, V-groove pine or spruce roof decking, Styrofoam insulation, felt and fiberglass seal-down shingles. Partitions and porch included. Wood-frame, double-glazed windows with screens in prefabricated sections; solid wood exterior doors prefabricated into sections; hollow-core interior doors. Second-floor decking is pine or spruce V-groove. Miscellaneous hardware, additional sealant, and exterior stain included. Model shown in photo is expanded Lakeside with optional full-length enclosed porch, breezeway, and garage.

Exterior Dimensions: 36×24'

Living Area: Upper   624 sq. ft.
            Lower   864 sq. ft.
            Total   1,488 sq. ft.

Price: $15,876 FOB Bellaire, MI

Price/sq. ft.: $10.67

Warranty: None, but Bellaire notes that it has had satisfied customers for more than 35 years.

Special Features: Balcony can handle four double beds dorm-style or can be partitioned. Options include floor system, triple-glazed windows, insulated walls, and additional roof insulation. Porch is constructed for easy screening or enclosure. Company estimates that about 80% of its sales are now for year-round rather than vacation homes. Bellaire has been in business for more than 35 years and delivers throughout the country.

Notes: Shipped on company trucks at $1.60 per mile. Only one load required. Three or four people needed for unloading; construction time about three weeks for crew of three. Bellaire says a large percentage of its homes are erected by purchaser, and "Any layman can do the job." Little caulking involved.

Price of Information: $2.50

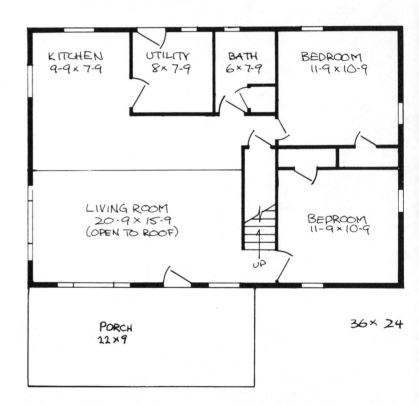

170

Lodge Logs Trapper

Price of Information Kit: $5

Type of Structure: One-story log partial shell package

Manufacturer: Lodge Logs by MacGregor
3200 Gowen Road
Boise, ID 83705
(208) 336-2450

Materials Provided: Package is for walls and interior partitions only. Logs are pretreated lodgepole pine, single tongue-and-groove system with fiberglass sill sealer and key blocks wrapped with fiberglass. Logs are air-dried, with a curing kerf to permit full exposure to air. Maximum log length is 10'; 6" to 10" logs with 3' bolts. Owner supplies floor, roof, doors, windows, and finish.

Exterior Dimensions: 20×26'

Living Area: 520 sq. ft. (1,040 sq. ft. with loft)

Price: $5,720.75 for 6" logs; $7,108.77 for 7" logs; $7,808.13 for 8" logs; FOB Boise, ID

Price/sq. ft.: $11 for 6" logs; $13.67 for 7" logs; $15 for 8" logs (without loft)

Warranty: Quality of materials guaranteed

Special Features: Some 52 models from which to choose, ranging from the Trapper to Custom series models up to $57,000. Because this is a partial shell package, customer will have additional investment in finishing. Options include trusses and Thermopane windows. Company has about 30 dealers; included in package price is supervision by dealer or factory rep for log-wall construction.

Notes: This model requires one flatbed (actually less; company tries to combine shipments to reduce cost, $1.30 per mile). Packaged in bundles 4×5×8-10' with fork lift recommended for unloading. Wall erection for three people estimated at about six to eight hours for this model.

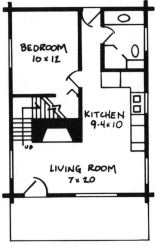

20 x 26

172

Hexagon 401

Type of Structure: Two-story panelized complete package

Manufacturer: Hexagon Housing Systems, Inc.
905 North Flood
Norman, OK 73069
(405) 321-2880

Materials Provided: Supporting structure consists of steel triangles composed of foundation sections, columns, beams, and plates. Exterior wall panels are prefabricated in sandwich form of a 2-3" styrofoam core with various finishes on each side, usually paneling inside and rough cedar, marble chips, painted plywood, granite, even gravel outside. Panels are 4×8', surrounded with wood splines for joining; panels sit in metal channels top and bottom which fasten to steel frame with self-tapping screws. Roofing is high-core steel. Package includes tile flooring, cabinets and vanities, range and hood, light fixtures, plumbing fixtures, heating and air-conditioning units, mirrors, acoustical tile and metal ceiling grid, 6" R-19 fiberglass insulation, fireplace, windows, doors, door jambs, battens and splines, and miscellaneous hardware.

Exterior Dimensions: 58'×56'; each side of hexagon is 16½'

Living Area: Upper   700 sq. ft.
Lower 2,100 sq. ft.
Total 2,800 sq. ft.

Price: Approximately $49,000 FOB Norman, OK

Price/sq. ft.: $17.50

Warranty: One year for materials and workmanship

Special Features: Floor plan shown here is downstairs only. Flexibility is the advantage with this system; a couple can start with one hexagon of about 700 sq. ft. and add three or four more in time. Wide range of finishes and appliances offered. Only limitation in floor plans is loca-tion of steel columns. Two men can handle any component in con-struction.

Notes: For single hexagon, a U-Haul or 24' truck is sufficient—larger models need 40' trucks. "With details and instructions, almost anyone can erect them," says Hexagon's literature. Tools normally needed are drill with magnetic sockets, circular saw with plywood blade and metal-cutting blade, level, and caulking gun. Units assembled with nuts, bolts, and screws for easy disassembly and rearrangement.

Price of Information Kit: $5

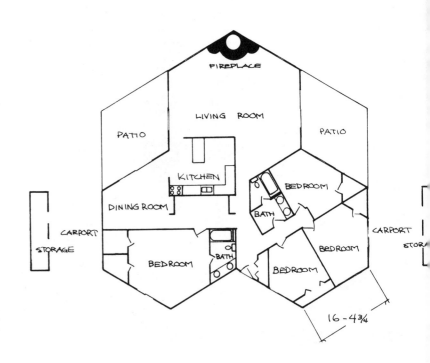

Habitat Colony Series Saltbox

Type of Structure: Two-story panelized shell and interior package

Manufacturer: Habitat
123 Elm Street
South Deerfield, MA 01373
(413) 665-4006

Materials Provided: Foundation of pressure-treated timbers (or owner may supply basement or slab foundation). Preassembled walls of 2×4s 16″ on center with 1″ styrofoam sheathing and prestained siding applied on-site. Roof of 2×8s 24″ on center with ½″ plywood sheathing and hand-split cedar shakes. Second-story floor system included, owner supplies first-story floor system; interior finish flooring of wide pine where specified. Double-hung, wood-frame windows with storms and screens; urethane-insulated steel exterior doors. Interior partitions; selected wood accent walls; ceiling beams of 6″×6″ hewn and axe-cut solid pine; prehung, raised-panel interior doors; prebuilt interior stairs; choice of four styles of kitchen cabinets. Not included: fiberglass insulation for walls, floor, roof; sheetrock; plumbing; wiring and fixtures; heating; floor tile or carpeting.

Exterior Dimensions: 44×22′

Living Area: Upper   384 sq. ft.
            Lower   648 sq. ft.
            Total   1,032 sq. ft.

Price: $19,006 FOB Lancaster City, PA

Price/sq. ft.: $18.42 for shell and interior package

Warranty: Manufacturers' warranties passed on to kit buyer

Special Features: This small-scale colonial home is compact without feeling cramped and uses high-quality materials throughout. Wainscoting and exposed beams add charm. Storage shed adds 48 sq. ft. of storage or can be converted to a solar room.

Energy Features: The compact size allows for low heating and cooling bills. With fiberglass insulation purchases to Habitat's specifications, walls are rated R-21 and upstairs ceiling R-30 (with 9″ insulation) or R-38 (with 12″ insulation).

Notes: Can be contractor- or owner-built. With five or six people, it can be unloaded from the flatbed truck within four hours.

Price of Information Kit: $1; $5 for package on Habitat's Colony Series, Timber-Kit homes and Passive Solar designs.

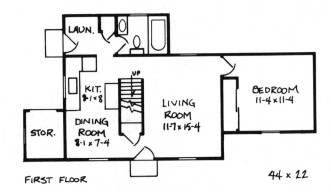

FIRST FLOOR   44 × 22

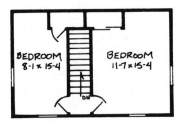

SECOND FLOOR

Yankee Barn Mark I-X

Type of Structure: Two-story post-and-beam panelized shell package

Manufacturer: Yankee Barn Homes
Drawer A
Grantham, NH 03753
(603) 863-4545

Materials Provided: Precut and numbered antique timber frame. Wall panels are sandwich construction of Douglas fir or eastern white pine plank exterior siding, two layers of solid urethane insulation, Douglas fir plank interior paneling bonded to supporting ribs, with insulating windows sealed into panel. Maximum panel size 8×13'; R-16 to R-24. Roof also panelized with Douglas fir plank ceiling, solid urethane, and plywood sheathing in panels 4×17' (owner supplies roof covering); completed roof rated R-31 to R-39. Floors are tongue-and-groove planks or panel system, beam seats on Lally columns and panelized units of tongue-and-groove Douglas fir, optional insulation and plywood subfloor on frame, maximum 4×18' size. Trim package for interior and exterior of rough-sawn heavyweight lumber. Completion kit includes solid-wood exterior doors, insulating sliding glass doors, miscellaneous trim and hardware. Also available is E-Wall package with unfinished interior wall panels for owner to insulate and finish to taste.

Exterior Dimensions: 48×26'

Living Area: Loft   84 sq. ft.
Upper   992 sq. ft.
Lower 1,248 sq. ft.
Total 2,324 sq. ft.

Price: $51,000 with plank floor system, FOB Grantham, NH

Price/sq. ft.: $21.94

Warranty: One year for materials

Special Features: Highly prefabricated home for ease of construction. A more complete kit than most, with attractive components. Also available are smaller designs, ells, foyers, and garages. Home packages range from about $20,000 to $51,000 in the complete panelized version, and in the new E series from about $26,000 to $50,000.

Energy Features: Factory-sealed wall panels include windows for less leakage and warping; rigid urethane insulation prepacked into panels throughout house.

Notes: With a crew of four, Yankee Barn claims, shell can be up in six days. Crane recommended. Yankee has its own crews and supervisors; recommends supervisor at $175 per day to oversee construction. Shipped on two flatbeds with crane, fork lift, or at least four persons for unloading.

Price of Information Kit: $5

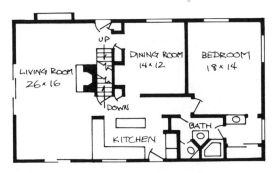

FIRST FLOOR

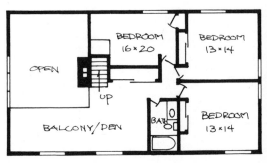

SECOND FLOOR

7 SEEING IT THROUGH

You've picked out your house kit, and you've sent the company some money. They are marking up your blueprints or cutting up your rafters, and you have a couple of months to prepare for delivery. Your work, at this point, will be just as critical to the success of the venture as theirs. They have to make the pieces fit, but you have to make the people fit. It's your responsibility to line up the bankers, insurance salesmen, power-company representatives, planners, laborers, craftsmen, inspectors, unloaders, and foundation experts, to get their various approvals and acceptances and to get them on your side. In your spare time, you have to get the foundation poured and the subfloor ready, the land cleared, and the sewer system chosen. The kit officials may help with all of this, or you may find that it is hard to reach them after the sale has been made. In any case, we have talked to company representatives and to people who have built kit houses, and they have offered the following bits of advice, admonition, and reminders of things that should not be overlooked. We take each situation in rough order of appearance.

Making the Deal

Kit prices may not be as fixed as they seem. Depending on how good business is at the moment, kit companies may be willing to give discounts for buyers in new areas, buyers who pay cash on delivery, or buyers who order kits in the slow season. We're not saying that you can always get a discount, but kit prices are not forged with untouchable sanctity. You might want to make a counteroffer and see what happens.

Kit prices are usually guaranteed from the signing of the contract until delivery 90 days later. If delivery does not occur during that period, due to no fault of the buyer, make sure the kit company will hold to its original price. You should not be penalized if the kit company falls prey to a lumber shortage or a labor strike.

When you work with the company to set a delivery schedule, don't forget the weather, especially if the delivery will take place in a rainy or snowy season. Rain and snow may make it impossible for the trucks to pull up to your site, and rain and snow have a habit of slowing up carpenters and ruining building materials left unprotected at the lot. If you plan to build in the winter, ask some local carpenters what delays and difficulties to expect. The best way to avoid all this is to build in the spring or in the summer.

Pinning Down the Contractor

If you plan to have a contractor build the whole house, or the shell of the house, you might as well pick one as early as possible. If you found one before you got this far, he was probably instrumental in helping you evaluate the various kits. A contractor will cost you money, perhaps 15% to 20% of the price of the completed house kit, and if you are dedicated to the adventure of self-building you will not want one. But there are advantages to working with contractors, as David Rohrs of Habitat explains:

"It sometimes helps a lot to get a contractor to work from the start of a house up to the erected shell. That is a good cut-off place, and you can do the rest of the work later. The contractor will help you out later. You have made an ally." The ally can help you make budgets for subcontractors, choose electricians and plumbers, get discounts on materials, deal with inspectors, and do a number of other things.

Choosing a contractor is just as tricky for a kit house as it is for a conventional house. You go to the bank for recommendations. You ask around. You take separate bids and compare them. The extra wrinkle with the house kit has to do with whether the contractor has ever built one of the kits before. If the choice gets down to a couple of contractors, pick the one who has had experience with the kit. He will be able to put it together more quickly than somebody who is not used to numbered timbers or color-coded beams. A couple of kit owners have told us that the contractors' lack of familiarity with kits cost them extra time and money.

The legalisms surrounding contractors are complex, and if you don't know the person involved, you may have to rely on him more than if you had an architect to design the house. An architect can look over a contractor's shoulder, but if you buy a kit, you will have to play that role. You may want a lawyer to draw up a contract. An article in the 1978 *House Beautiful* plans book suggests some legal language that will help ensure that contractors live up to their bargain:

— A specific completion date, followed by the phrase "time is of the essence" (i.e., June 10, 1978, time of the essence). "In court," said Long Island attorney Peter Feilbogen, "that phrase means the completion date is more than a joke."

—Include the following phrase "Workmanship must be the best known to the trade . . ."

—Include a guaranteed maximum price for the completed job, and a schedule of payments.

—Include this clause: "Final payment will be withheld until the contractor releases a proof of payment from major suppliers and all subcontractors."

The point of all this is to pin the contractor down as much as possible on schedules, prices, and workmanship. Legal language is a good line of last defense, and you will want to find out how typical contracts are drawn up in your area. But the best defense against shoddy work or careless contracting is to give out as little advance money as possible. If you have a mortgage, the bank will have its own stringent payment schedule. If you don't have a mortgage, you might learn from the way a bank does it. Payments should be given on work that has been completed. Some money should be held out until building has been inspected and approved. Contractors should show all bills for materials. Under no circumstances should you advance a contractor a substantial portion of the money before work has begun. Many people have been burned in recent years by contractors who required such large advances, never did the work, and then declared bankruptcy. Find out what protection you have in the local licensing or bonding of contractors.

Getting the Permits

You have to get permits, no matter what kind of house you build. Contractors usually handle it, but if you are building yourself then you will be responsible. The local planning board or building department will tell you what you need. If you aren't hooking up to city sewers, you will have to do a percolation test on your land to see if it is suitable for a septic tank. You will also have to bring in some plans. If the kit company does not send your personalized blueprints until a few days before the truck arrives, you'd better get the permits on some general description of the kit. You don't want to become the owner of an expensive pile of lumber that you aren't allowed to put together.

As we have mentioned, kit companies generally build homes that meet or exceed structural or energy standards in all the major codes. As long as your city or town abides by some recognized code, there should be no problem. Once in a while, a local official will balk at some detail in a kit plan. John Odegaard, a director of Mayhill Homes, a Georgia company that does not sell to do-it-yourselfers, gave us two recent examples. The building powers in two North Carolina towns questioned Mayhill kits. One of them didn't like a particular roof-truss arrangement, and the other thought that interior walls should be cross-braced. When the Mayhill technical department came up with the numbers to support their plans, the local planners issued the permits.

Usually there will not be a flat turndown (unless a *type* of house is prohibited), and if there is a disagreement, the house kit company may be a helpful ally. You should find out from the company how much support they will give if a snag occurs. Shelter-Kit says it is the "obligation of the buyer to observe applicable local codes and regulations." Other companies share that obligation. If there is some question about permits, make sure the kit company will refund your money if the permits are ultimately denied.

There is the secondary issue of whether the various codes and regulations allow do-it-yourself building. Find out. This is also the time to ask about building inspection. Is there anything about the house kit plan that would complicate the inspection process? Some kits arrive with the plumbing and wiring already installed in the finished wall, so the inspector could not very well look it over without tearing up the house.

Getting the Money

Most house kit companies do not lend money directly—in fact, they are emphatic about getting all *their* money a day or two after the kit is delivered. The bank deal is up to you. The kit people may help you convince skeptical bank officers that a dome won't peel away like a dandelion in a high wind. But you have to persuade the bank to finance the kit.

If you are working with a contractor, getting a loan on a kit won't be any harder than getting a loan on a plan for a stick house. But if you are doing it yourself, you may have to try a number of different banks. Not all banks are willing to open their checkbooks to amateur carpenters, and they wince at the prospect of repossessing the leaky lean-to that was supposed to be a ranch house. "The banks," says David Rohrs of Habitat, "will look at an individual hard."

Before you get a loan, banks need to be convinced of the intelligence of the plan, satisfied that you have the skills to build it, and assured that your cost estimates are reasonable. These hurdles may be infuriating, but they can also be beneficial. It is easy enough to float away on the fantasy of a kit house, and if a banker can still see you, then you are still near enough to solid ground.

The requirements of banks have limited the market on a lot of house kits to people who have money. Banks will peer harder into a person's own safe-deposit box if they think the house itself might not be such wonderful collateral. The recent increase in popularity of house kits has done something to change this situation. Banks are starting to loosen up a bit. FHA and VA have approved a wide variety of kits, including log-cabin kits, for their financing, and if you can convince the FHA, then you have automatically convinced a lot of building-code people and bankers. A few successful do-it-yourself stories can do wonders for the cash flow into house kits.

When banks do lend for house kits, they may not offer a traditional long-term mortgage right away. It is an important thing to know when you are shopping around for money. "If an individual is building himself," says David Rohrs, "he may have to work through interim financing until the shell is completed." There are various ways this interim financing can be secured. Sometimes the land itself becomes the collateral, and sometimes other assets are involved. The bank wants to protect its money until it can send somebody around to check the finished house. You get the short-term loan, or swing loan, and then convert to a regular mortgage when the house passes bank inspection.

These short-terms loans make the economic calculations for house kits a little different than for regular construction. So does the large initial payment that you must make to the kit company. In regular construction loans, the bank pays the contractor in small installments, so during the first couple of weeks of work you still haven't borrowed that much. With kits, you borrow more right from the beginning to pay for the materials package. The payment schedule for Real Log Homes is typical: a 10% deposit, 40% due 30 days before the shipping date, and 50% on delivery of the house kit. (Real Logs also accepts a 10% deposit and 90% on delivery if it has a letter from the lending bank confirming the 90% payment.)

If you are borrowing a lot of money, you can't be too relaxed about getting the house kit built. You may have to start paying interest 30 to 90 days before the trucks arrive, or at least from the day the trucks arrive. Some kit companies will schedule the deliveries for early morning, so you can be hammering by the time you have checked over the inventory and the trucks are out of sight. The virtue of house kits is that they go up quickly. High interest rates make it imperative that you take advantage of that virtue.

The interest factor should be included in your calculations on the total cost of a house kit project. Do the savings on speedy construction offset the cost of borrowing a huge chunk of money early in the building process? It is a question you can discuss with the banks and with the kit company. Charles Firestone, who put up a barn kit in Virginia, says he spent over $6,000 on interest in the first year, and that was before the barn was even completed.

Firestone's experience in general is worth recounting. He and his wife, both of whom make a good income, went to their bank with the idea of being the contractors and building the barn home. Because of the way the kit company operated, they did not have floor plans or blueprints to show the banks. "Maybe I should have asked them for more money, but it was already a very large mortgage on a set of plans that didn't exist," Firestone says. His hefty cost overruns were not covered by the mortgage. Firestone suggests that when people contract out their own house kits, they should get a very accurate estimate of the final costs. Banks sometimes don't know enough to help them.

Firestone's other suggestion is that kit buyers investigate the banks' regular payment schedules. Most banks have rigid timetables for releasing money to contractors. They pay out so much when the foundation is completed, so much for the stud walls, etc. Since house kits are put together in an unusual manner, the pace of construction doesn't always conform to the bank's timetable. "One day this bank person came out," Firestone says. "A payment was due to the carpenters. It wasn't the usual bank representative — the regular person was away on a trip or something. This guy didn't want to release the money. He said, 'How can I give you the money for phase three when some phase-one things aren't completed?' I reminded him it was a kit house, and that the phases didn't fit his schedule. I told him some phase-five things were already done. It took some argument, but I finally convinced him to make the payment, although it was reduced from the amount that should have been paid at that time."

There are four things to remember when you go to a bank with a house kit: (1) Just because one banks says no, it doesn't mean that they will all say no. Many kit builders are rejected at one place and welcomed in the next, so you have to keep trying. (2) Take note of the interest you have to pay on the money to buy the kit, the interest on the swing loan, and how these charges affect the cost of the house. (3) Get enough money. If you finance on a fantasy budget, you will have to scramble and scrape to pay for the overruns. (4) Make sure the bank understands that payment schedules for a kit house will be different from schedules for a conventionally built house.

If you are lucky enough to find a kit company that finances its own homes, you won't have to bother with any of this. Miles Homes makes an offer that is worth repeating. Through its unusual plan, Miles reaches many people who could not otherwise afford kits, or any other type of home. The company requires no down payment. Miles sends along the materials for the kit, and also gives you a short-term construction loan, backed with your land as collateral. You have **two** years to build the house, during which time you are making modest payments to Miles. After 22 months, the principal is due. By this time, if things have proceeded according to plan, you will have completed the house and can qualify for a traditional mortgage. The bank pays off Miles, and you continue paying the bank. Miles gets its money, and you get a house with no down payment and small early payments during construction.

Preparing the Site

The local building department and any number of popular books on construction can help you here. The costs of surveying, sewer lines and septic tanks, grading, tree removal, and backhoe work should be included in your building budget. The extra concerns of kit houses include making sure the trucks can get in and out, providing a space to unload materials near the foundation area, and finding some protective covering for kit materials.

To ensure that work can begin immediately after the truck is unloaded, you should talk to the utility company about installing a temporary electric pole. This is another job that contractors normally do, and contractors usually have their own temporary power poles that they carry from site to site. When you do the work yourself, you may have to buy the components of such a pole. Ask your utility company what is required before they will make a hookup. Make sure you allow plenty of time for the power company to complete the connection.

You may also need to buy some extra lumber to use in bracing and leveling the walls of the kit. You can ask the kit company what is required here.

The most critical part of your preparation may be getting insurance to cover the subcontractors and workers on the job. Contractors have their own insurance, but if the electrician falls off the deck and you are the contractor, you may be liable for damages. It is advisable to get such insurance before any worker shows up at the site.

Hiring the People

Even if you do most of the work yourself, you will probably need a foundation crew, a plumber, an electrician, a roofer, and maybe a carpenter or two to help when you get bogged down. Unless you know them personally, it is best to do some investigating before hiring a subcontractor. Unfortunately, the yellow pages do not carry useful headings like "craftsman" or "punk." You can get recommendations from the bank, from a contractor, or from the kit company. The kit people often have lists of men and women who have worked on kit houses and know the idiosyncrasies of kits. You can call the owners of those houses to find out what kind of a job was done by the subs. If the choice can be reduced to two crews who will do the work for about the same price, pick the crew that has already worked on one of the kits.

Once you have some names, you will want to get estimates. How you phrase an estimate will have an effect on its accuracy. If subcontractors know they will not be bound to the estimate, then they will tend to shave things a bit to lighten your mood before hiring. After the job is done, they can woefully present the real bill, with some solemn allusions to "cost overruns." You want them to be as specific as possible. You can pin them down by telling them to put a maximum cost on the job and that if they exceed that cost, they will have to pay the difference. Some subs will not work with this kind of fixed-ceiling estimate, but others will.

The other trick with estimates is to ensure that the various subs are figuring for the same kinds of materials. All of them will claim to work "to code," but a building code may offer a lot of leeway between expensive and cheap alternatives. Plumbing pipes, for instance, can either be made from copper or from PVC plastic. Some codes allow both. A plumber who is figuring for PVC will obviously have a much lower materials cost than one who figures for copper. There are similar variations in the size and quality of electrical conduit, fuse or breaker boxes, and switches; gutter materials, flashing, and shingles; concrete mixes and reinforcing bars. The best way to produce a reliable materials estimate is to get a list of exact specifications and have your subs work from that. If you are not capable of drawing up such a list, the kit company or a contractor might help you. It would probably be worth a few dollars to pay a contractor to write up the specs. Not only will you get a more accurate estimate, you will earn the respect of the subs, who will begin to believe you know what you are talking about. A good spec sheet will help with quality control while the work is in progress. If the specs call for 2,500-pound concrete, it's not going to be as easy for a foundation person to put in 2,000. If the specs are vague or nonexistent, the subs can skimp on quality to increase their profit.

Estimates should be divided up between materials and labor. Most of the subs will know where to get discounts on materials, and they may be willing to share those discounts with you. It happens that way with a lot of do-it-yourself contractors, but it won't happen unless you ask. The best time to ask is during the estimate phase, when the subs know they are in competition for the job.

The labor part of the estimate will not be as accurate as the materials part. We have heard from a few kit buyers who say that the subcontractors took longer than expected because they were not accustomed to working with the kit or with a particular kind of house. Electricians and plumbers may view their skills with a certain poetic license, so it's best to pin them down on their actual experience. If one labor estimate seems much lower than all the others, it should be viewed with some caution. You will ensure accuracy by holding the subcontractors to a maximum labor cost for the job.

There are all sorts of legal remedies that can be pursued if the work is done sloppily or incorrectly, but they are rarely worth the time, trouble, or your lawyer's fees. The important thing is preventive contracting. If you are not working on the house yourself, frequent site visits will show the subcontractors that you are serious about the work. The way you handle the money will be even more convincing. The subs should be held to a specific time schedule, with money paid as work is completed. Some subcontractors sign up for a lot of jobs and then juggle their excuses to keep all their bosses only moderately dissatisfied. These delays can damage a building budget, especially if you are paying high interest on the money. Most subs will not work under penalty clauses, but you can set up an incentive system, offering a bonus if work is completed on schedule. But the main thing is to stagger payments so the money never gets too far ahead of the job. Subcontractors may require some advances for the purchase of materials, but they should not require advances on labor. We strongly suggest that you hold back at least some of the final payment until the job has been inspected and approved. Subcontractors who were easy to find before a job begins have a way of becoming hard to find once the job has ended, especially if there are problems.

Usually, all the subcontractors will take care of their own withholding taxes, workmen's compensation, etc. You should ask if those items are covered, and also find out if they are insured. Even if they are, we strongly advise that you carry your own insurance.

Making a Foundation

Most house kits offer an option on foundations. You can put the house on a full basement, a crawl space, piers or pilings, or a concrete slab. House kits require that the foundation can be completed by the buyer. In most instances, the foundation must be poured and the joists and subfloor constructed before the kit materials arrive at the site.

The kit people will help you choose the right foundation for your situation, but here are the advantages and disadvantages of the four options:

PIERS OR PILINGS. You drive poles into the ground, with a certain spacing in between them. Then you set the house on top of the poles. The poles can be made from concrete, concrete blocks or wood. If they are concrete, they are called piers; if they are wood, they are called pilings. The main advantage of piers or pilings is the cost. Since the foundation does not have to be continuous, you save on materials and you save on labor.

The disadvantage is that the entire underside of the house is exposed to the cold air. There is no basement in which to run plumbing pipes or furnace ducts, and the heat loss through the first floor can be tremendous. Piers or pilings can also settle unevenly, giving a house a slight tilt to one side or another.

SLABS. Slabs are basically large beds of concrete, poured onto the ground or poured over a plastic barrier. Slabs can take the place of the entire joist and subfloor structure of a house, so the savings there can make up for the expense of the concrete. The problem with slabs is that they don't sit far off the ground. Houses built on slabs can suffer from heat-loss problems, moisture problems, and insect problems. They are frequently used in southern climates with concrete-block houses, but for wooden houses, slabs are rarely advisable as foundations.

CRAWL SPACE. Crawl-space foundations are built like basements, but without headroom. They provide a protected area under a house, but don't leave enough room to walk around or stand up. Some houses have partial basements and partial crawl spaces, to take advantage of the roominess of the former and the economies of the latter.

Crawl spaces are usually made from concrete block, set on a concrete footing that runs around the perimeter of the house. The footing is at least 12 inches below the frost line, to ensure that it won't crack. The joist system is built on top of a wooden sill that covers the top row of concrete blocks.

Recent energy studies show that a lot of heat can be lost through and around crawl spaces. It is important to insulate crawl-space walls with some sort of Styrofoam or polystyrene plastic barrier. The polystyrene should extend all the way down to the footing of the foundation.

BASEMENTS. Basements are the most useful and the most expensive foundations. They are made from concrete blocks, or from poured concrete. Basements are continuous, which means that they distribute the weight of the house evenly all along the perimeter. An insulated basement adds usable floor space to a building, and provides a convenient refuge for the furnace, water heater, water pump and storage tank, etc. A Princeton energy project discovered that up to 25% of the total furnace heat may be lost to a basement, so unless yours is insulated, it may be advisable to put the furnace somewhere in the living area of the house. If the basement is insulated, the heat that is given off by the furnace itself can rise up through the first floor.

All of these foundations share one important characteristic: If anything goes wrong with them, you have troubles up the line. A foundation that is not in square can start a chain of compensations that starts with the joists and doesn't end until the shingles are applied. Such a small error at the bottom can cause headaches and extra hours devoted to fitting, fudging, and figuring.

A foundation error in a conventionally built house may be more tolerable than one in a kit house. In a conventional house, the carpenter has to cut each joist and each rafter anyway, so it is possible for him to take an inch here or there to compensate for mistakes. Since the framing members in a kit house are precut, there are no fudge factors. This is one way in which building a kit house requires more skill than building a stick house. Tolerances are critical.

All this leads to a piece of advice: Don't do the foundation yourself unless you are an expert. Pouring concrete and making footings are not pleasant jobs, especially for people who are looking forward to a poetic rendezvous with wood. Foundation work is gritty and critical. You have to know about concrete, steel reinforcing bars, frost factors, waterproofing, drainage, settling, block laying, squaring, and leveling. Not only should you get a professional to do the work, you should get a kit representative or salesman to come over and measure the perimeter and check out the job before the kit is dispatched on the truck. This precaution can save a great deal of money, wasted time, and embarrassment. In fact, many kit companies *require* such an on-site inspection; most are quite firm about what kind of foundation is necessary.

In hiring the foundation crew, you should go through the same procedures as you follow for hiring plumbers or electricians. Check with the building department to see if an inspection is necessary before you proceed upward from the concrete blocks.

Meeting the Truck

Have more than enough people on hand to help with unloading. Go through each piece carefully, to make sure that nothing is damaged and nothing is missing. This is the best time to lodge a complaint, not when the truck leaves. Kit companies require payment on delivery, but some companies give you a couple of days to go through everything before you send the money. This grace period is worth bargaining for. Some buyers hold back a little of the money, say 5%, for a week or two until they are certain that everything is in order. Kit companies do not like this practice.

If there is some difficulty at the site that impedes the unloading, it could cost you money. Dave Rohrs, at Habitat, says that trucks customarily charge $20 an hour. If a truck is delayed on your property because you aren't prepared, you will have to pay the overtime.

You might want to insist that the trucks arrive in the morning. It will give you and your friends enough daylight to unload the materials, and it will ensure that the truck will not be kept overnight if delays do occur.

Building the Kit

It's all in the manual. If you're still confused at this point, sell your tools and buy a condominium.

INDEX